W9-BLP-366

FINDING THE FLAVORS WE LOST

ALSO BY PATRIC KUH

An Available Man: A Novel

The Last Days of Haute Cuisine: The Coming of Age of American Restaurants

FINDING THE FLAVORS WE LOST

FROM BREAD TO BOURBON, HOW ARTISANS RECLAIMED AMERICAN FOOD

PATRIC KUH

ecco
An Imprint of HarperCollins*Publishers*

HarperCollins books may be purchased for educational, business, or sales promotional use. For information please e-mail the Special Markets Department at SPsales@harpercollins.com.

FIRST EDITION

Designed by Shannon Nicole Plunkett

Library of Congress Cataloging-in-Publication Data has been applied for.

ISBN 978-0-06-221954-1

16 17 18 19 20 OV/RRD 10 9 8 7 6 5 4 3 2 1

FOR MY SISTER,
MICAELA

CONTENTS

PART ONE

TO LAND 3

TO CRAFT 21

TO PLACE 41

TO MARKET 59

TO TABLE 75

PART TWO

TO DISTILL 103

TO BAKE 119

TO BREW 133

PART THREE

TO ESSENCE 155

TO WORK 177

TO SCALE 199

TO HARVEST 217

TO TELL 235

CODA 253

ACKNOWLEDGMENTS 275

FINDING THE FLAVORS WE LOST

PART ONE

TO LAND

128TH AVENUE, FENNVILLE, MICHIGAN
OCTOBER 28, 1978

It was the day of Alyce Birchenough and Doug Wolbert's wedding. The home-brewed root beer was chilling, some friends were setting up their bluegrass instruments on the porch, and others were hauling kegs of beer when a local farmer pulled into the rutted driveway. Hitched to his truck was a trailer holding a single cow. She was to be Doug's present to Alyce, the animal that would help them to revitalize the thirty-three acres of depleted soil they'd settled on in western Michigan. Doug hadn't nailed down the delivery date and, attuned to the casual ways of Fennville, he reasoned this just happened to be a lull when the farmer had time. Still, he was more than a little surprised to learn the cow hadn't been milked. Was that customary? There she was, in the trailer, her udder tight, obviously in pain. Alyce and Doug exchanged a glance. Alyce immediately understood that this was Doug's gift. But today? Had he forgotten to tell the farmer that this was precisely the date not to bring the cow? Then again, maybe he had and this was a rural statement: you wanted country living, well, here you go.

Their relationship with the locals had been much on Alyce's mind in the few months since she had moved to Fennville. They lived in the town, a small farming community near the eastern shore of Lake Michigan with an active 4-H youth agriculture club, sport leagues, and churches, but they were not of it. Instead of buying their staples from the local grocery stores, they got them in bulk from a Chicago co-op whose truck intermittently undertook the four-hour drive to service a ragtag group of communes. Alyce winced at how this could be interpreted as rejecting the community, but she had come to the land to be, in a word, de-Twinkiefied, and she was actively pursuing that when she stood by the vehicle's rusty tailgate using her own scales to measure out whole wheat flour, brown rice, and chickpeas.

Less than a mile from their homestead, Fenn Valley Vineyards had opened a few years previously and was selling natural fruit wines, but the townspeople mainly thought of it as a curiosity. Alyce and her friends were an even greater one. Dungarees, beards, homespun vests, loose handmade dresses cut from patterns, dinged-up hats; they looked like farmers all right—from another century. Still, there was no animosity; perhaps there was even a hint of admiration at their sense of thrift. The newcomers' vehicles—a fleet of sputtering, backfiring decal-covered vans—weren't going to show up anyone's pickup. The home-canned tomatoes that might bring sunshine to a winter table were not that far from what was going on in surrounding kitchens. When a farmer whose peaceable Guernsey was being picked on by the Holsteins in the herd decided to sell her, he put a notice in the local paper and Doug responded. Whatever differences existed had not prevented a sale.

Alyce had gotten into the habit of extrapolating great significance from little evidence. Those times when, driving over the rutted dirt byway that linked their farm to asphalted roads, she received a country greeting—finger lifted off steering wheel from an oncoming car—she could be happy for days. They were being accepted. Now, standing before her, was a farmer pointing out that the cow he had with him hadn't been milked and she understood that instead of a taunt, this was rural showmanship, a way of displaying for them just how much milk the cow could produce.

She could contemplate those subtleties later. There wasn't time for rumination. She saw the cow pawing in its trailer in obvious discomfort. Alyce and Doug had readied a shed across from the house and she led the cow toward it while Doug finished up with the farmer. Rushing back across the driveway and into the kitchen, she found an enamel pail and, remembering quickly the steps of milking she'd studied in anticipation of this moment in Carla Emery's *The Encyclopedia of Country Living,* she raced back out to the animal. *Grab high on teat near udder,* the first of Emery's three drawings instructed. Applying the gentle pressure that Emery described, while all the time squeezing to force the milk downward, Alyce started the repetitive up-and-down movements. Who was this person she was becoming? She wasn't some kind of pioneer woman who would have been practiced in these farmstead skills. She was someone who less than a decade earlier had been a science major celebrating the first Earth Day with much strumming of guitars on the campus of Southeastern Louisiana University. Now she was taking possession of a cow on her wedding day. Sounds often accompany life's turning points—the creak of a door closing in a place we won't return to, a newborn's cry—and Alyce felt there was something of that to the hissing the milk made as it hit the bottom of the enamel pail.

BY THE FOLLOWING WEEK DOUG and Alyce had gotten used to the rhythms of milking and named their Guernsey Cindy Lou. The occasional mooing they'd hear from the shed or the sight of her swishing her tail nose down in a tuft of grass contributed a sense of permanence to their new life. It was your typical 1970s homestead: a modest structure, a scrawny garden, some sheds, a three-seater outhouse, and only five workable acres. The poor quality of the land—heavy with sand that blew from the deforested edges of Lake Michigan—was part of the reason Doug had been able to buy the property. His first task had been to clear it of the car shells and debris that the previous owners had left. Soon after, he'd set up a sawmill that ran from power he took off a rotating tractor belt, and he often accepted lumber in payment for transforming a log into a stack of boards. It was a simple life, punctuated with occasional parties that gathered the local back-to-the-land community in one setting.

Alyce had attended one of those gatherings when, unsure where her life was going, she'd first visited Michigan. A New Orleans native, Alyce had moved to Connecticut soon after graduating from college and worked for a time as a nutritionist at Yale–New Haven Hospital. It had been a disappointing experience; the doctors barely took her suggestions seriously. The nods her nutrition and wellness recommendations received were only a maddening form of dismissal—they were never included in the instructions the doctors jotted down. While others might have made their peace with the situation, Alyce saw no need to cede as important a point. She'd left, traveled, and eventually visited a college roommate friend living in the town of Holland, an artistically inclined town on Lake Michigan, fifteen miles north of Fennville.

The two friends had driven down to one of Doug's parties and, over the course of a few weeks, Alyce fell in love with the rangy Wolbert. His farm and its land needed a lot of work—the brush pushed right against the back of the house—but she admired how he was going at it, slowly and alone, armed with little more than a deep appreciation for self-sufficiency, something she, too, shared. Many of the improvements he'd made to the hodgepodge house had an endearingly off-kilter quality—pulling a chain in the kitchen let a jet of fresh air in through a roof trap—and over repeated visits she started to picture herself joining him in his quiet life. It didn't take much convincing to get her to stay—still, she had standards. The three-seater outhouse had to become indoor plumbing. In a climate with only 120 frost-free days a year it didn't seem to Doug such an outlandish demand.

Perhaps a little comfort was perfectly fitting, even deserved. Though there was a long tradition of sects rejecting the trappings of cities, disengagement did not necessarily induce hardship. Rappites, Mennonites, Shakers, Perfectionists—all managed to pursue transcendent goals while enjoying their share of temporal pleasures in the form of breads, dried fruits, vinegars, whiskeys, and hams. Charles Nordhoff's 1875 study of these communities put it succinctly. "I have found it generally true," he wrote after years of research, "that the members of communistic societies take life easy."

If anything, modern times had taken some of that simplicity

away. No previous landward-looking generations had been more curiously studied or mercilessly skewered than those of the 1960s and 1970s. News coverage dripped with condescension. One expected it from *Fortune* and *Time,* but it bore a particular sting when written by contemporaries who cast the urge to move to the land as a tiresome set piece. Instead of finding solidarity in the pages of *Working Papers for a New Society*—a short-lived quarterly aimed against the status quo—one found mistrust. There's almost a prosecutorial thoroughness to Rosabeth Moss Kanter's 1974 description of a typical urban commune interior. Typically, such "wayside inns on the route to the country" consisted of "windows full of plants, batches of drying herbs, tattered oriental carpets, clutter of furniture, pets, smells of incense and marijuana, old barn boards and rough wood shelves in the kitchen holding mason jars of organic grains." In the same publication, in a 1973 issue, Andrew Kopkind chose to narrow the focus to a single Vermont commune, the aptly named Mayday Farms, which "occupied a comparative middle ground between the austere and the amenable in style, between the revolutionary and the psychedelic in politics, between the farmers and the hangers-out, between the lower and the upper in class background of the members." With such a jumble of elements, it was no wonder few communes lasted.

As homesteaders, Alyce and Doug were spared the travails that accompanied communal living. They were landowners involved in at least a nominal form of farming with a few chickens and, now with their cow, even animal husbandry. Already, two weeks after Cindy Lou's arrival, Alyce could not keep up with the volume of milk she could produce. Containers filled the fridge. She tried churning butter, added Hansen's rennet tablets she bought at the store to make custardlike junket pudding. Going one step further, she even tried her hand at making cheese. Her food sciences training made the process of cheesemaking appear simple. The initial acidification lowered the pH to counter any lurking pathogens, culturing established the desired microbes, a measure of rennet allowed for the separation of cultures and liquid, and . . . nothing came out. Nothing that they would want to eat at least. It was time to turn to Carla Emery again.

Started in 1970 with an ad in *Organic Gardening* magazine, Emery's standard guide to surviving on the land was for its first seven printings a series of three-hole-punched mimeographed sheets that spoke of hard-won experiences. In the late '60s, she'd moved to an almost deserted stretch in northern Idaho and soon found herself in what she later described as "a tremendous out-migration from cities to country." With time she became the voice of authority on every conceivable aspect of self-sufficiency. Despite passages on serene activities such as candle making, to read Emery's *Encyclopedia* is to be reminded of the realness of the homesteading enterprise. "You could use a .22 but a 30-30 or bigger is surer," she says in her typically thorough manner on the subject of gun gauges used for cow slaughtering. "Try very hard to get it right with the first shot." Cheesemaking could hardly be considered a challenge after that.

And yet, despite Emery's clear instructions, the cheeses Alyce tried to make were not cooperating. To help her, Doug bought her a Wagner's Gourmet Home Cheesery, an ill-conceived beginner's kit popular at the time. The press where the cheese was molded was a coffee-can-size plastic cylinder with a threaded rod running down the center. Pressure was applied with plates and wing nuts on either end. The problem was that when the cheese was set and the rod drawn out, it left a long, hollow passage behind. A moist, dark, air-filled hole, precisely the kind of spot where pathogens could establish a home base. The cheese didn't mature because it rotted first. It was enough to make Alyce question the whole undertaking. Country greetings were nice, but Alyce and Doug weren't country people. What was this undertaking of theirs?

If she had been a pioneer woman, the knowledge of cheesemaking would have been there, passed on from a mother or a family tradition. Dairy work had always been the work of women, a way to sustain the household in the long periods between payment for crops. We can appreciate just how demanding that work was in the 1898 diary a Mormon woman left where, with a brevity that itself speaks of fatigue, she delineates the daily tasks.

> SAT 17TH — Churned six pounds of butter
>
> MON 19TH — Churned eight pounds butter, done up my work, pickled some beans
>
> FRI 23RD — I help milk all the time. John [son] and me do all the milking. We milk nine cows but the churning is the hardest part.
>
> TUES 4TH — I churned eight pounds butter, made pies, cooked dinner. I have to turn and rub my cheese every day. The wind still blows. I read and knit some.

The wind still blows . . . it is an almost haunting allusion to the one thing that didn't demand something from her. An impossibly hard life, yes, but the recipes were internalized; there was no fooling with wing nuts and starter kits. The knowledge was there. Alyce had chosen the land but had bypassed a tradition of knowledge. Each batch of cheese she made she fed to the chickens.

TODAY WE ARE SO ACCUSTOMED to having immediate access to sources of information that it is good to be reminded of how people in another time organized their network of knowledge. It was the age of the self-addressed stamped envelope, or SASE. The nation was buzzing with requests for information and mailed replies with answers. You sent away for knowledge, included a SASE, and eventually, dropping through your letterbox, back knowledge came. Emery was but one among many. The hand-lettered recipes that Mollie Katzen developed at the Moosewood Restaurant, located in a former school converted to miniaturized community in Ithaca, New York, were bound in spirals, turned into a book, and soon after publication in 1977 were so popular she was spending as much time at the post office as in the kitchen. While a student at the University of Virginia in 1972 Charlie Papazian had taken up home brewing; by 1978, he was compiling the mimeographed sheets he gave to students taking his brewing class at Boulder's Community Free School—sheets that eventually became a beer- and mead-making primer. It could be ordered, by mail, if you sent the stamps.

What these works share—apart from accompanying drawings in an R. Crumb style of shaggy-haired characters applying themselves to vats, stills, pickling pots, and hand-cranked churners—is a beginner's spirit. The author had figured out just that much more than the reader (and in the case of one John Barleycorn who instructed on hooch making from Willits, California, that included federal law) and was pretty much passing that knowledge on. In food-related activities where expertise had always been closely guarded by guilds, laws, and industry groups, these sometimes ditzy publications represented a sea change. If you were a beginner, the expert was suddenly on your side.

The success of Ricki Carroll, founder of the New England Cheesemaking Supply Company, exemplifies these newly forged expert ranks. Over the course of books, workshops, and presentations, Carroll's smiling face has given thousands of enthusiasts hope that their gallon of milk will turn into something edible derived from curds. She has told the story of her beginnings so often that it seems almost intrinsic. After moving to Ashfield, a small town located between the Connecticut River and the Berkshire Mountains of western Massachusetts in the early '70s, she, her husband, Bob, and the assorted people they lived with in a rambling, peak-gabled house needed to find a source of income. A neighbor had convinced them there was money in goats, and Ricki started to try to learn feta making, in the hope of someday selling it to local co-ops and restaurants.

A bright and enthusiastic woman, Carroll understood they'd have to improve the quality of cheesemaking to have any hope of keeping their house. Determined not to return to teaching art in New Jersey, she and Bob wrote letters to twenty Washington, D.C., embassies asking whether they knew of anyone who taught home cheesemaking. They got one reply, through the British embassy, from a farm in Bristol in southern England. The farmer there would teach them how to make, among others, the hard white cheese that was a favorite of Welsh miners, Caerphilly. Bob decided that there might be a market of other enthusiasts looking for supplies, and just before their departure to England, he took a $9 ad in the back of *Dairy Goat Journal*. It read: "For a catalogue of cheesemaking supplies send 25 cents." The ad was one square inch. They didn't think more about it until two months later when they returned and were barely able to open their P.O. box at the Ashfield post office because it was so full.

Bob and Ricki set about filling the requests for supplies that weren't always easy to get hold of. They'd buy rennet tablets from Hansen's and drive north to the American Fiber and Finishing mill in Colrain to buy large bolts of cheesecloth and, at home, cut it into two-yard lengths. By late 1978, a person writing to the New England Cheesemaking Supply Company could receive in a single bag all she could possibly need, including a basic culture, a mold, and a recipe. Later, the kit included a blueprint for a cheese press and a plastic curd cutter that arrived embedded amid scrunched-up pages of the *Ashfield News*.

IN FENNVILLE, ALYCE'S EDUCATION HAD progressed significantly in the four years since buying Cindy Lou. Determined to use only unpasteurized milk, she had to work fast to use up the whole amount of milk each day, acidifying her milk as quickly as possible to create the acid medium hostile to pathogens. Though she had an ideal style—something that could be cut for a snack or sliced for a sandwich—the means of reaching it amounted to a short list of improvisational skills. Her scientific background allowed her to isolate and propagate the acidifying cultures she used. Doug deemed barbells the ideal weight to press the liquid out from the freshly formed wheels, and, by replacing the torn screen door of a pie chest with a new one, he had transformed it into a cheese cave. To coat each cheese, they mixed melted crayons with the puttylike paraffin Alyce used to give an airtight seal to jars of preserved vegetables. Perhaps it all did amount to a system, but one that could certainly be improved. When Alyce heard of New England Cheesemaking Supply through a friend, she quickly sent away for individual parts of Ricki's kit. What immediately impressed them—particularly Doug with his technical bent—was the blueprint for the wall-mounted press. Here, finally, were people who understood the notion of a fulcrum as a means to exert pressure. Compared to the barbells they were using or even to the cylinder and wing-nut combo of the Wagner's kit, it felt like a real advance.

BACK IN ASHFIELD, THE CARROLLS were becoming leaders in the new system of information exchange. In 1981, Ricki and Bob would start publishing books they thought would interest others. Their first one was a translation of a primer on goat cheese making originally written by the Mont-Laurier Benedictine nuns in Ontario. But even before that, by March 1981 they had started a newsletter grandly called *The Cheese Press*—and put an issue in with every order. Ricki's kitchen soon became the location for one-day workshops. Among the first people to sign up were Marjorie Susman and Marian Pollack, a couple who had driven on back roads from Leyden, a nearby Massachusetts town. They found themselves in Ricki's kitchen with its double sink, round table, and four-burner household stove. By the end of the class they had formed a fresh Caerphilly—and had drunk enough of Bob's dandelion wine to consider it a step taken toward their own unclear future.

Marian and Marjorie had met a few years earlier in the crowded upstairs office of a Greenfield group training to campaign for passage of the pending Equal Rights Amendment legislation. The two women inhabited different realms of the town. Marjorie was attending Greenfield Community College and had applied to the Stockbridge School of Agriculture, part of the state's university system. Marian worked as a family therapist for Franklin Community Action Corporation and, though a strong and determined woman, was growing hopeless about any change she could effect. Recently one of her charges—a teenage boy she'd thought of as one of her more promising cases—had pulled a knife on his mother. She'd never mistaken rural for idyllic, but there was something depressing about the slow grind of problems to solve and flare-ups she rushed to subdue.

Despite Marjorie's plans to study in Massachusetts, the women didn't see themselves living there. Marian had grown up watching her mother keep a Victory garden in Fairlawn, New Jersey, so it wasn't as if she was accustomed to the dew-dropped prairie. But she did want a life in the country, and Massachusetts, with its increasing density and prohibitive land prices, didn't seem like the place that would provide it. When Marjorie graduated from Stockbridge, they were ready for a change. In 1980, they placed an ad in the *New England Farmer* looking for positions on a working farm. When they received an offer to be herdspeople in Morrisville, near Stowe, Vermont, a small town north of Montpelier, they took it immediately.

The move wasn't exactly fraught—from Massachusetts to Vermont could hardly be called momentous—but it did establish them as outsiders within a traditional hierarchy. For Vermonters, the population they referred to as "summer people" or, between themselves, "flatlanders" could be counted on to arrive after Memorial Day, peak in number by high summer, and begin an exodus to the city that got under way before Labor Day and was complete by the first frost.

But another kind of migration was taking place, of local people leaving the land. The steps were just as gradual yet far more final. Those summer people would be back; farmers who left wouldn't. A 1964 study by the University of Vermont found that fully three-quarters of those people in the central part of the state—towns such as Bethel, not far from Stowe—who gave up farming remained in the town. When asked what they had gained, those who responded answered a greater income and freedom from a 365-days-a-year job. Asked what they'd given up, over one-fifth said there'd been no loss at all. But 14 percent pointed toward something plangent, answering they'd forfeited what was intangible, "a way of life."

In truth, that way of life had hardly ever exerted enough pull to keep Vermonters in Vermont. The lore of the Green Mountain state may well be due to nostalgic Vermonters. Life was hard in those dark valleys, and the opening of even early means of transportation created ways of escaping it just as compelling as accessing it. In the late eighteenth century, turnpikes had taken Vermonters to the New England seaports. By 1838, a Vermonter who could cover the cost of a cent-and-a-half-per-mile passage could embark on a steamboat in Whitehall and through a network of packet boats and steamers that plied the Champlain and Erie Canals be transported via Troy and Buffalo, ever westward toward Detroit and Chicago.

"Peddlers, printers, apprentices, bankrupts, and the rest were all only minor variations on the main theme," wrote Lewis D. Stilwell in his 1948 history, *Migration from Vermont.* "The bulk of the emigrants were simply surplus farmers of the common sort who were going out where land was cheaper and richer, and the winters were not quite so hard. The very ease and naturalness of the movement makes it seem the more convincing and inevitable. There was little or no agitation about it. No one was drumming up settlers in a craze or crusade. Vermont was like one of its own hillside springs, continually overflowing to the enrichment of the lowlands."

If there was a point of satisfaction for newcomers like Marjorie and Marian, it was that they were putting their boots into the state's soil. They understood that unless you had been born in the bounds of the township, spent your time around the box stove, and walked and sugared the woodlot, you wouldn't really be considered "from" Vermont. But they also sensed a certain recognition, perhaps even a form of acceptance, from the famously terse locals, for those newcomers who took on the challenge of living there. Though the job in Morrisville hadn't proved ideal—ironically, they suspected their employers had plans to sell the land for a housing development—they learned the kind of skills that were invaluable if one wanted to be a real farmer. Soon Marjorie was backing up a tractor with a manure spreader like a pro, and they felt confident enough that in March 1981 they answered an ad for help in the *Burlington Free Press* and were soon milking a herd near Middlebury.

The job they were offered came with a dilapidated farmhouse that stood off an unpaved road. It was here that the two women now lived. When they stopped working at mealtimes or at the end of the day, they found themselves looking out the window from the kitchen at the fenced expanse of the Champlain Valley that stretched toward the foothills of the Green Mountains. The couple soon had formed a real attachment to this land. Here they could settle. A year after arriving, using a no-interest loan from the Lotta Crabtree Trust, left by the famous turn-of-the-century actress (known as "the nation's darling") whose love of animals had led to a large bequest to what became the University of Massachusetts, they bought fifteen acres.

"We thought we'd get goats," recalls Marian, "because that's what girls do." But it was only a passing thought. They bought their first cow—Sultana—in May from a local farmer, and it was Sultana's calf—Timothy—that made them give the farm a name so they could register it with the American Jersey Cattle Association. They picked the name of a spider, Orb Weaver, whose translucent webs they often found spun between the stalks of their growing garden. At each stage in their settlement they knew they were being watched by locals. This was Vermont—the land of dilettante farmers who tried it and moved on. But all they'd see in Marjorie and Marian were two hardworking farmers they had to respect.

They planted a garden and took their vegetables into the co-op and restaurants in Middlebury, the local college town, to sell. They soon were milking their employer's herd on a nearby farm. They took on the responsibilities of raising heifers in their barn. With each achievement they gained unspoken approval. Marjorie and Marian bought their first cheese vat at auction from a dairy that was going out of business. They didn't know how to maneuver such a large piece of equipment to the shed where they'd be making the cheese, but a local farmer, a man who wore red suspenders over flannel shirts and was the very image of old Vermont, offered to help them move it in his truck. Vermont style: no charge.

Back in Ashfield, Ricki Carroll had become a clearinghouse for information. "If it was going on in the cheese world," she recalls, "we knew about it." Early in 1982, Frank Kosikowski, noted professor of food science at Cornell University, contacted Ricki to tell her about a meeting he was convening at the campus in Ithaca, New York. Kos, as he was known by students, had long been fascinated with the colorful variety of dairy products throughout the world. His major work, *Cheese and Fermented*

Milk Foods, initially published in 1977, was a seminal investigation into the wide-reaching world of cheese. The flyleaf was a drawing he did of sheep gathered round a water hole in Roquefort; the varied content of the text ranged from methods of drying yak curds in the Himalayas to the finer points of milk fermentation in industrial-sized cheesemaking plants. That was a subject he knew something about; he had spent decades teaching people who went on to work in the largest operations. But he had also realized that there was a parallel small-scale cheesemaking world active in the United States—and that attention should be paid to it. His idea was to form a society where all cheesemakers—homesteaders and academics—could learn from one another, and he wanted Ricki to put the word out to the legions of cheesemakers, aficionados, and budding artisans whom the Cornell Dairy Sciences program would ordinarily not be able to reach.

Ricki placed a notice in the June-July 1982 issue of her journal, and in Michigan Alyce Birchenough read it. Among other things, it announced that a professor by the name of Kosikowski was inspired by the American Wine Society to consider forming a similar organization for cheese. By the time the next issue appeared, the dates had been set for an initial seminar: June 5 through June 7, 1983, at Cornell's Robert Purcell Union. Dining and reasonably priced lodging facilities were mentioned. Ithaca and the Finger Lakes region were highlighted as attractions. A constitution would be drafted; suggestions on subjects for seminars could be sent to Dr. Kosikowski—care of Ricki and Bob's Ashfield P.O. box.

Ordinarily, Alyce would have shied away from such a gathering. In her mind, "big cheese" represented a turn toward homogenization. But did her cheese have any distinct flavor either? While she no longer made the "chicken cheese" she'd started with, she was achieving nothing more than bland spheres she shared with friends. She wanted to get better. She knew she needed insider knowledge, hands-on learning experience to improve. Since the pamphlet she received after writing away for information made no mention of the minimum herd size required to attend the gathering, she decided she'd go. Why not?

And so, in the summer of 1983, she and Doug drove in their truck to Ithaca to attend the inaugural meeting of the American Cheese Society. In the three days of seminars, they received invaluable information from experts. The inaugural conference truly was a cross section of individuals; one presenter spoke about creating huge volumes of standardized cheese loafs, another about the possibility of making brandy from whey. Just as important for Alyce and Doug, they met many others who had also come from small farms around the country. In fact, their number was a validation of Kosikowski's inkling about the breadth of the new cheese world. At the end of each day, the campus parking lot became a gathering spot where stories, tips, and addresses were exchanged around parked vehicles. (Alyce and Doug slept under their truck's camper shell.) No one had quite realized how many others were on this journey of learning a craft that was foreign to their backgrounds, or how many of those who had moved to the land were now embarked on the next phase of their chosen life: how to learn the skills that would allow them to stay there. Alyce finally felt as if there *was* a movement—and that she belonged.

TO CRAFT

In the summer of 1988, Nancy Silverton dragged four five-gallon containers full of active yeast starter into a warehouse space in Los Angeles. She was determined to make naturally leavened bread. She felt that she had all that she needed to do so. She'd carefully nursed the starter from an initial mix of organic grapes, flour, and water until it had become this slurry, not unlike pancake mix. She had a handful of bread recipes from a course in French baking she'd recently taken at the Lenôtre baking school in Paris, and she had a vision of starting a bakery in L.A., where none like it existed. That none of these recipes called for the use of the slurry was not a matter of concern to her. The bakers in Paris used compressed yeast since making naturally leavened bread was a technique that had died out in France, save for a few exceptions. Those bakers who maintained the tradition called themselves *artisans*—a word that unless you sold hand-punched leather belts at craft fairs had little meaning in the United States. Yet, this was the bread that Silverton was determined to make, a bread where slow fermentation allowed microorganisms to both leaven and ferment the dough, thus creating a certain acidity, a tang that bread lovers consider to be the very essence of bread.

In fact none of these recipes would prove even remotely workable. She'd had no doubt that it would be difficult—she'd never so much as taken a science course and the bubbling, frothy mass in the buckets was organic chemistry in action. Using a natural starter was an all-out commitment; the stuff was as demanding as a baby without granting the benefit of an occasional smile. From the start, everything seemed to be going wrong: the bread didn't rise, often took on weird shapes when it did, and no two batches looked alike. Then there was the matter of taste. The tang that was supposed to be the payoff for all this work was completely undetectable in her dull crumb.

She worried that the slurry had not been fed enough of the raw flour it needed to remain active. Or had she overdone it and put in too much? Was it too thin or too thick, too hot or too cold? Her troubles didn't end there. Flour was flour; at least she'd thought so when she'd started. But no, that was not the case. The French recipes she was armed with depended on French flour. Now she had to start investigating American flours and modifying the recipes—the one thing she'd thought was set. When she paused for long enough to look around the space she worked in with its fluorescent light, sink, mixer, and deck oven not even designed for baking bread, she realized that in truth she had nothing. She wasn't so much in a borrowed kitchen as she was in a crucible. If she was to become a baker, it would have to happen here.

The space had been loaned to Silverton by a friend who'd used it for a catering business. The low-slung structure was unprepossessing. It was located on the lower stretches of Robertson Boulevard, close to, though not quite in, the shadow of an elevated section of the 10 Freeway that crosses Los Angeles, and, eventually, the United States. With bars on all the windows, walls sheathed in corrugated metal, and a thin coil of barbed wire running around the parking lot, all that could have been added to the dismal scene was the presence of a junkyard dog. Inside stretched a succession of kitchens used by small businesses. One made muffins, another pasta sauce, a third prepared perishable goods sold at Trader Joe's. Though their activity was constant, and weathered delivery vans seemed to be always pulling in and out, the location was monastically quiet compared to the activity she'd been fleeing when she grabbed her buckets of starter and left the construction site that would eventually become Campanile restaurant.

That growing structure represented its own difficulties. And, unlike the *levain,* they weren't of the kind a scoop of flour could fix. Scaffolding had to be put up with much clatter, power tools must be run incessantly, and if a mason was in the habit of working to the strains of blasted KROQ, then blasted KROQ it must be. What had she been thinking to consider this a suitable venue to try the most ancient form of baking? But the restaurant's creation had to continue also. She knew that. Silverton and her husband, Mark Peel, had long dreamed of running their own kitchen. Now they were doing just that, reconfiguring a classic 1920s structure on La Brea Avenue into the business of their dreams. The building had belonged to the mother-in-law of one of Charlie Chaplin's brides—the tiled central fountain and the sweeping arches spoke of Los Angeles's fascination with Moorish architecture, while a verdigris bell tower rising above it struck an Italian note. It was a mishmash of cultures with a definite Old Hollywood feel. When the couple had first toured the property, it housed a miscellany of businesses ranging from the offices of a community newspaper to a small music school to a massage parlor that had a threadbare runner guiding customers to its door. But Silverton and Peel saw it as the perfect location for the combination restaurant and working bakery that they wanted to build, a place where Southern Californian and Mediterranean traditions could meet on a plate. Now scaffolding was draping the inside walls, dust covered everything, and the infernal noise of jackhammers made it hard to talk, let alone bake. When Silverton's friend suggested she borrow her kitchen space, Silverton loaded four buckets of her finicky starter into the back of her Saab and fled.

Despite its potential to be a distraction to Silverton, creating the restaurant represented the culmination of a certain journey, one that symbolized the heyday of California cuisine. If Berkeley's countercultural openness marked the initial phase of the style's development, then Los Angeles with its mass density and promotional possibilities marked the culminating one. In Northern California, the style's inception could be narrowed to one restaurant, Chez Panisse, just a short walk from the University of California's, Berkeley campus; in Southern California, the fulfillment of the vision could be traced to two establishments, Wolfgang Puck's Spago perched over Sunset Boulevard in West Hollywood and Michael's, a breezy place known for its garden a few blocks from the palisades that overlooked Santa Monica State Beach.

Peel had started in the kitchen at Michael's in 1979, and Silverton arrived soon after. They didn't yet know each other, but they'd both been attracted by the restaurant's reputation as the center of a culinary revolution. Peel was the more seasoned of the two, having already worked at both Ma Maison and Chez Panisse. Silverton was a newbie, a graduate of Le Cordon Bleu in London; she was not able to even get a kitchen job at Michael's but was instead given the task of entering customer orders on restaurant owner Michael McCarty's new computer system. She barely understood computers, but it didn't matter; she was in, working at the restaurant that was getting press coverage throughout the nation.

With a patio that spoke of outdoor living, peach-colored walls that were a nod to the soft-hued ones of the Beverly Hills Hotel, and its garden setting, McCarty's restaurant was a whole new formulation of Southern Californian glamour. Here, waiters wore button-down oxford shirts; nibbles were spread with chèvre; and culinary ingredients such as baby vegetables, fresh local prawns, mâche—that indispensable ingredient of all nouvelle cuisine salads—and butter sauces were all kept spinning around the maypole of a mesquite-fired grill.

France represented a backdrop for the restaurant, almost the context through which the Michael's kitchen articulated the lyricism of California. They were not the first to attempt the feat of one place speaking through another. Already in 1899 Charles Ranhofer of Delmonico's in New York had brought the tang of the rocky shores and redwoods to his menu in cream of green corn à la Mendocino, and Chez Panisse had been grappling with the feat of envisioning Amador County zinfandel for Beaujolais since soon after opening in 1971. Like no other restaurant had done, McCarty's laid out the themes of the argument; the culinary directness drew on references as broad as fisherman stew cioppinos and backyard dome-topped Webers; the umbrellas and shimmering light of the setting alluded to French reserve and proportion. Today it is easy to think of the tension between France and California that rippled beneath Michael's as some kind of turf war that occurred so long ago that it is unimportant. It seems quaint, in our age of borderless inspiration, where Thai and Mexican influences can happily coexist on a menu, to think two titans had to face off. For California to assert itself then was an openly defiant act. And yet it had to be done. All the characteristics of modern dining—vital, rustic, local, authentic—stem from a battle in which traditional notions of good taste (and what was in better taste than haute cuisine?) were challenged by an upstart group of American cooks.

Silverton got to see some of these skirmishes firsthand once she made it into the pastry department (she'd never quite gotten the hang of those computers). *Department* was perhaps too grand a word. The area where desserts were made consisted of a small marble-topped table and a double-stacked oven squeezed between the stoves and the expediting pass the waiters picked up the dishes from. Away from the hot station, she and head pastry chef Jimmy Brinkley grappled with the conflicting principles of elegance and directness that the battle for the future of dining hinged on. In pastry, the differences were particularly pronounced. The classic dessert had to be ethereal and sublime, a wisp of a wisp—but was that really a fitting end to a meal that highlighted the robust flavors of the grill?

Silverton had still not developed the confidently rustic style she would later become known for. She remembers being impressed by Brinkley's version of crème brûlée, which was cooked on the stovetop, then poured into a large puff pastry shell and browned under a broiler. Presented on the dessert table by the entrance—a classically elegant gesture—it seemed to represent something sophisticated, but the crackle of the caramelized crust spoke of something far more fundamental than the usual haute cuisine offering. Her own desserts would also have to find that balance between rusticity and refinement. Under McCarty's guidance, she started to understand that many classic desserts—glazed and perfect though they may have been—had an overaerated, mousselike quality. She also learned that pastry could have intense flavor. "I eventually understood," she recalls, "that there was a difference between a good French pastry and a good pastry based on French technique."

SILVERTON'S PASTRY-MAKING JOURNEY HAD A certain structure, a focus that would ultimately manifest itself in the desserts of Campanile. However, the journey that would make her a baker—an artisan despite herself—was serendipitous, nothing more than an offshoot of her work in kitchens. As a dorm cook at Sonoma State University, her cooking was completely vegetarian. The occasional loaves she'd bake from the natural food cookbooks she used all had an earnest solidity, revealing more about ethos than craft. At Michael's, the bread had been Pioneer sourdough, the same bread she'd grown up eating when her mother would buy two loaves, or "flutes," to a bag at the Gelson's supermarket in Encino. Even when she and Peel became part of the opening crew of Spago in 1982, her duties didn't at first include making bread—just lots of desserts.

After three years spent at Spago, Peel and Silverton had a reputation in the food world for having played a central role in the development of California cuisine. In 1985, restaurateur Warner LeRoy tapped them to revitalize Maxwell's Plum, a massive Upper East Side restaurant with a Tiffany ceiling and a reputation as a pickup joint that had lost some of the pizzazz it had been famous for in the 1970s. But instead of energizing the business, the differences between East and West Coast approaches were exacerbated when the free-flowing nature of kitchens that Peel and Silverton thrived in came up against the hard reality of an owner who insisted on bringing back dishes from the old menu when regulars asked for them—and a long-established unionized staff. "You'd have a cook cut a cake with the same knife he'd cut smoked salmon with and you couldn't do anything about it," Peel recalls. Without much prompting, the restaurant they wanted to create started to take shape in their minds. They returned to L.A. after a disastrous seven-month stint in New York.

Puck was glad to see them return and offered them jobs at Spago, even though he understood that by then Peel and Silverton were planning their own restaurant. Enough of a working system had been established at Spago for Puck to now focus on the in-house bread program he'd always wanted to create. It wasn't overwhelmingly ambitious. One of the two breads, a multigrain loaf, was made from a mix a friend of Puck's had discovered in Germany. The second bread, however, an olive and thyme loaf, drew on a fascination with the Mediterranean that had been lit in Puck while he was a young cook at L'Oustau de Baumanière, a perennially Michelin-starred restaurant in the craggy medieval village of Les-Baux-de-Provence. In Puck's mind it would be a crusty and intense loaf, boosted with black olives and leavened if possible by natural starter. The bread program that Silverton and kitchen manager David Gingrass created was as good as could be done within a restaurant that didn't have a separate area for baking. Silverton secured some starter from her old friend Brinkley, who was now working at the Sign of the Dove in New York, and she and Gingrass daily set about making Spago's breads.

"I don't remember that the olive bread had all the characteristics that we now have at La Brea Bakery," she recalls. Still, those hours spent watching the starter, looking over the loaves, timing the bakes, and gauging the doneness were the beginning of a fascination. "With pastry there was no reason for it not to come out the same every single time unless I wasn't concentrating," she says. Naturally leavened bread was different. And how. By the time she had reached the caterer's kitchen under the freeway, she was starting to find out just how different the two disciplines were. Everything imaginable could affect the process. The moisture of sea air on a foggy day might speed up the fermentation; dry heat tended to bloat the dough, making it flabby and hard to handle; turning on the air conditioner created a dry layer, which inhibited the rising. And yet nothing deterred her from this petulant mass that was the very foundation of a commercial bakery. "I was set on doing a starter using wild yeast. I felt that was the challenge in baking and that was the challenge I wanted to face."

She laughs now at the disdain she had for commercial yeast then, calling it nothing less than a crutch. The easy rise, the miracle that required only tepid water to be activated, how lame to take such an easy approach when you could harvest the real thing starting with grape skins! She wasn't alone. Steve Sullivan was attempting the same thing at Acme Bread Company in Berkeley. Heck, the Boudin Bakery in San Francisco had been saving a little of yesterday's batch for today's bake since 1849. Clearly she had kindred spirits who were interested in creating dough the right way.

It seems a tad harsh to demean a process that was once considered a godsend and had taken centuries to master. For millennia, fermentation had been started by adding something already fermenting to a bigger batch of what was to be fermented. Yogurt in Mesopotamia, beer in Saxon England, vinegar in medieval France—even sour mash bourbon in today's Kentucky—were created using this method, kick-starting a process whose inner workings were barely understood. Sheer human inquisitiveness had long since figured out that a yeasty ladle of froth in the baking trough helped bread rise. But there the knowledge stopped. Even as recently as 1803, an English treatise on bread making states that to get enough yeast to run a commercial bakery, one must have "a boiler, cooler, vats and all the apparatus that would be necessary for a small brewery." If snipping open an envelope of Fleischmann's yeast was a crutch, then so be it.

People rejecting the long tradition of scientific advancement that eventually tamed the natural fermentation process were engaged in a far larger argument than how to get bread to rise. They were making a statement, taking a stand for that flavor-intensifying period of time that yeast, once isolated, had made obsolete. In bread baking, time equals flavor. The slow process of dough's maturation allows for a complexity to develop to which the heat of the oven lends a billowing crumb and crisp, dark crust. Commercial yeast had sped up this process so that rising dough could blow through stages that had previously taken hours. Flavor was lost; time was gained. Putting time back into the equation was striking a blow against the very principles of efficiency that an industry demanded. To search for flavor was to accept inefficiency as a business model and that, in the world of industrialized food, was a rejection indeed.

The truth is that Silverton was launching into the most ancient form of baking at a time when even yeast-using, efficient, industrialized bakeries were becoming endangered. A few blocks away from the kitchen where Silverton was working stood the large, abandoned structure where the Helms Bakery had existed. The business opened in 1931—its walls adorned with a cluster of rings, a nod to the Olympics the city would host the next year—and soon after, its paneled trucks were delivering fresh loaves directly to people's homes. By the time the business closed in 1969, that fleet of neighborhood bread trucks, emblazoned with a hand-painted logo that read DAILY AT YOUR DOOR, was a relic from another time. *Twice a week to your supermarket shelf* was more like it.

America had changed enormously in the postwar period; the distances foods traveled to their point of sale were vast; the economies of scale these operations required became ever more pressing. People could once know the baking schedule of a plant by the aroma of warm rolls wafting over their neighborhoods. That was gone by the late '60s. Bread no longer came from local and specific places such as the Helms Bakery on Venice Boulevard in Culver City, California. It issued, like just about everything else, from distant loading docks where pallets of food were loaded into idling eighteen-wheelers during the dark of night.

Countering that anonymity was merely a vague concern of Silverton's at the time. She had more pressing matters. Just getting the bread to come out right was far more immediate. Though technically she was trying to make loaves from soured dough, she was not trying to make what is popularly referred to as sourdough bread. That bread, fermented with a particular strain of bacteria, is associated with San Francisco and has a distinct sourness. It represented a different tradition from the one she was pursuing. American sourdough originated with "forty-niners" and gold-panning Klondike miners who, when necessary, slept with their starters in bowls under blankets to save them from succumbing to the cold. Silverton's starter was *levain,* the stuff of lore, a word capable of evoking a France of small villages and hand-shaped, hearth-baked loaves, a word so rich with associations it exerted close to a primal pull on the French mind.

But how strong was that tradition of *levain* even in France? In *Good Bread Is Back,* his book on the recent return to high-quality breads in France, food historian Steven L. Kaplan paints a bleak picture of decreasing quality. In the Gironde district, to cite but one of his statistics, the number of artisanal bakeries had decreased from 1,080 in 1955 to 655 thirty years later. What happened in those intervening decades was a pure example of social changes affecting gustatory traditions. A highly Americanized version of economic growth caused the country to shed its rituals. Increasing wealth made bread less of a staple; with the nation trending toward middle-class values, bakers no longer wanted to work the deathly exhausting hours of their trade (Marx called the pale, flour-dusted baker boys of his time "white miners"). All this occurred as food technicians devised a method to ship frozen, preshaped dough from a central factory to outlets that would conduct the baking. The result: those limp baguettes that seem more steamed than baked that can now be bought in any big-box supermarket in France.

The movement back to good bread began with Pierre Poilâne, a Parisian baker who single-handedly preserved the hearth-baked boule, a large country loaf with a dark crust that appears frosted by the flour of the proofing bowls in which it rises. Throughout the 1950s, just as France was beginning its period of economic progress, Poilâne ran something like a counterinsurgency from his Left Bank workshop. Kaplan remembers meeting him in the 1960s—Poilâne wore a beret and a gray artisan's smock and had a taste for Sancerre. Pierre's son, Lionel, also donned the traditional baker's robe but did so while seeking new markets, and perhaps even a new role, for celebrated breads like the two-kilogram *miche*. With his guidance, darkly baked, flour-encrusted Poilâne loaves became newly chic. Today's generation of bakers, men like Éric Kayser and Dominique Saibron, have brought the tradition further, modernizing it completely with licensing arrangements, consulting partnerships, and multiple stores. The American artisanal baker can look at such a logical progression with only envy. There were no men in smocks to learn from here. No words like *pain au levain* or *pain au feu de bois* that reeled customers in with their associative connections. Were there even any customers here? The little family bakeries had mostly disappeared; so had many of the big ones. Those people who wanted to learn the true craft of baking were on their own. There was no guild, no craftsperson, no tradition to allude to, no vocabulary to hitch one's efforts to, and no one to ask.

Books offered little encouragement to Silverton. In *Beard on Bread,* James Beard gives instructions for salt-rising bread (the term comes from the farmhouse tradition of keeping the starter on a bed of warmed rock salt, which retains the heat) in less than enthusiastic terms. "It is unpredictable," he writes before sending those who wish to try it forth. "If it works, fine; if it doesn't, forget it . . . Good luck!" Even a paean to authentic baking pulled its punches. Bernard Clayton Jr.'s 1978 classic, *The Breads of France and How to Bake Them in Your Own Kitchen,* includes in its ingredient list for a Poilâne boule—the very definition of a naturally leavened bread—one package of dry yeast.

"The hard thing about bread is the waiting," says Silverton, recalling that period. "The baguette I was making had a twenty-four-hour turn time, so everything looked great, the dough felt great, I shaped it the way I'd learned to shape it in France and I'd put it in the refrigerator, and then twenty-four hours later I'd take it out of the oven and it would be just like a rock." Batch after batch like this. Such was the lonely pursuit of the self-taught artisan, a painstaking process on which the artisanal movement in this country would be founded.

Each time she took another ruined batch out, Silverton felt like she had failed. Perhaps efficiency was a worthwhile goal, something to be lauded instead of sneered at. Who was she to criticize anyone who could produce thousands of loaves a day? Slotted into toasters at breakfast, taken out of lunch boxes, buttered in diners, whatever its purpose, plain mass-produced bread gave plenty of pleasure. Was this some kind of smug moral superiority? A way to separate herself from others? No one had set out to eliminate flavor from bread. They'd cut out the time traditional leavening required, and flavor had simply vanished with it. But those bakers were home now. Time had bought them a life. She was in a warehouse with the clock ticking toward midnight with this endless rotation of loaves that never quite came out right. This was a moment of truth. She now understood baking intellectually but had still not mastered the craft. It took another batch, another journey toward knowledge that began with a twenty-five-pound bag of flour emptied into the bowl of the Hobart mixer. The more exhausted she was, the clearer she glimpsed a basic truth: artisanship had nothing to do with a lifestyle—it was a personal quest that one decidedly took on.

Silverton's learning was gradual, but the pressure from the restaurant's imminent opening was constant. Silverton and Peel had set out to open both bakery and restaurant simultaneously.

Soon enough they'd realized this was impossible, and they got a big break when they were able to secure separate permits from the city of Los Angeles for the businesses. Only the bakery had to be open before the end of the year to fulfill the terms. Slowly, she felt she was getting there. In October 1988 Silverton moved out of the rental kitchen and into what would become La Brea Bakery. The restaurant was unfinished; the bakery little more than a brick shell. There were basic pieces of equipment she didn't have: a retarder whose cool chambers keep dough from swelling in hot bakeries, a steam-injected oven that boosts the "oven rise" and helps create the dark crust of the classic boule. Even so, she left the Robertson Boulevard kitchen with recipes for five breads, enough she felt confident to open a store with.

The country white loaf was the very definition of the naturally leavened, hearth-baked loaf she'd set out to craft. Though the sourdough baguette was far from perfected—she worried about its shelf life—she felt like she needed one since that was what customers would expect to find in a European-influenced bakery. The rosemary olive oil loaf was a hearty variation of the country white one, as indeed was the walnut bread—both used the same dough but added different ingredients. Initially, Silverton had been wary of selling a whole wheat bread because of its health-food connotations. Eventually she'd become fascinated by the challenge of this heavier flour and was proud of the final loaf that rounded out her list.

In her effort to perfect these breads, she had left out many of the more mundane concerns of production. Worrying about flours, testing proofing times, hadn't left her time to think about the layout of equipment. She hadn't even thought of where the racks would go when they came out of the oven at a sizzling 450 degrees. Touch them and you'd be left scored like a steak on a grill. Now she had to quickly come up with solutions. The room that was going to be Campanile's dining room was a jumble of carpenter's trestle tables and scaffolding. The racks couldn't be placed there. The only solution she could come up with was to wheel them out to the parking lot and let them cool there. Who would see it? Surely one of the benefits of being a baker was the certainty that Health Department inspectors weren't driving around at 3 A.M.

What worried her more than these practicalities was that she still felt like an interloper in the hallowed halls of artisanal baking. She sometimes caught herself thinking she had no right to be there. Who was she? A girl from Encino who'd grown up eating sourdough sold at Gelson's supermarket. A sometime cook who'd gotten the baking bug and figured out five recipes that sort of worked. That was it. Hardly someone related to flour-dusted artisans who gracefully wielded long-handled peels. Their heritage was supposed to be her career? Come on.

What happened then was one of the strange confluences of eras that the artisanal movement was built on. Silverton met Izzy Cohen, a baker from that older tradition that was being plowed under by mechanization. Cohen was past seventy by the time chance brought them together, and his last job had been decorating cakes at a bakery in the Beverlywood neighborhood of Los Angeles—ironically, not far from the loaner kitchen where Silverton had struggled. There wasn't much of a market for an artisan trained to make bagels and naturally leavened rye. Meeting Silverton, Cohen found a baker who was trying to revive techniques he'd thought forever lost. Both risked becoming culinary anachronisms; Cohen by mixes that required only water,

Silverton by quick-bake versions boosted with massive amounts of yeast. "He walked into the bakery one day and asked me if my breads came from mix," she recalls. "I told him no, they were made from scratch." The concept of this more traditional approach resounded with Cohen. Soon he was a constant presence, giving Silverton the guidance of his half century of baking. The baker who still felt unworthy of the tradition she was embracing was being instructed by one who had lost a tradition and felt cast aside.

Cohen's first lesson was in bagel making, something that Silverton had long wanted to learn. His initial lessons were practical; philosophical knowledge followed shortly afterward. "His knowledge of starter was of a rye starter because he was a Jewish baker. He taught me how I could change the starters just by changing the flour I was feeding them. And he taught me how to maintain a starter at optimum potency. Starter has a life, kind of like a graph. When you first feed it, it's at the bottom of the bell curve. It is strongest eight hours afterward and then it starts to fade again. I redid all of my timing so I was able to use my starter at the top of the bell curve, and that made a huge difference in the consistency of my bread."

She was starting to feel like a baker. She was acquiring confidence. "He taught me bakers' percentages," Silverton says, recalling Cohen, who died a few years after the bakery opened. "That's really important because it is how bakers write recipes. They don't use the cups and ounces I was writing down then." Cohen also taught her the nuances of scoring the bread before it went into the oven. There wasn't a standardized slash; you had to gauge the depth of the cut by the type of flour (rye rises less than white flour and a deep cut would never open up properly).

Even the degree of proofing had to be taken into account since a slash could either billow or sit on a loaf like a scar. She had to learn to gauge these things by touch and in a matter of seconds. But with time Silverton was doing the slashing confidently, using a real curved *lame,* or blade, instead of the shaving blades she'd bought at the local drugstore. She took pride in carrying the *lame* between her teeth—just like a baker.

Between these lessons there was time for banter. "He used to laugh at me," says Silverton, who dedicated *Nancy Silverton's Breads from La Brea Bakery* to Cohen and to Acme Bread founder, Steve Sullivan. "The expensive things I'd put into the bread, the virgin olive oil and fresh herbs in the olive bread, he said no one would notice." Here was Izzy Cohen, stooped from scoliosis, and Nancy Silverton, the nervous beginner. She was part of the food revolution; he'd grown up in the Great Depression. A person who had the luxury of choosing to go into baking contrasted with a person who'd had no choice. Yet he was her teacher; he knew all the tricks and all the techniques. They were an unlikely pair, but here, finally, was knowledge being passed on.

What he gave her, in short, was the confidence that she could step into the role of baker. She had done battle with the natural starter and tamed it sufficiently so that the buckets of unstable, living sludge she'd started out with could become the foundation of a bakery. The phase when every variable had seemed unfixed was ending; the one when it was becoming a business was almost starting. To comply with the separate permits that Peel and Silverton had been granted, the bakery was required to open before the end of the year. All the other things she still had to learn would have to happen on the job. Now they needed a customer.

By late December 1988 Silverton was making mixes daily.

Varying the ingredients slightly, she perfected each bread's characteristics; using her bakers' percentages, she increased the daily bake to several dozen loaves a day. After it had been formed, the bread simply waited, rising in the silence that she had come to love. She waited with it. She still had no idea of the coming craziness that would descend when she and her workers would be bagging bread in the parking lot before dawn. Those early days when she'd realize she hadn't even ordered paper bags to put the bread in, those entire months when she'd be sleeping in hours snatched from the day and would wake up with crazy baker's dreams, anxious stuff about forgotten batches, when her whole life seemed like one more ingredient in the baking cycle. That was in the future. Now the first batch for the public rose in floured baskets. She let the loaves rise, and then scored them assuredly with a single movement, knowing instinctually by now how deep the slash should be. She placed the loaves in the oven, and when they were done, she took them out and let them cool. Manfred Krankl, the manager of the still unopened restaurant, put the loaves in a paper bag and walked them three blocks north on La Brea Avenue to City Café, where they were signed and paid for. In that restaurant's dining room, waiters were decorating the tables for New Year's Eve dinner. It was the last day of 1988. The first order had been taken, the first delivery made.

TO PLACE

Student furniture is the same everywhere, passed down through generations of shared housing until a couch ends up on a porch or in a backyard. The covering may be worn and the legs may invariably wobble, but these pieces retain something of the conversations that took place on their deeply furrowed cushions. Ken Wright and Allen Holstein wiled away the hours on one of these very couches—in their case, an old settee with busted springs—at the University of Kentucky in Lexington, discussing the nuances of red Burgundy.

Wright had discovered the wines of the region (a realm beyond the quasi-sophisticated pink-hued Lancers that constituted a splurge for the roommates) while waiting tables at a nearby high-end white-tablecloth restaurant, the Fig Tree. During horse races at Keeneland, the local track, or even eighty miles away at Louisville's Churchill Downs, the wines sold themselves. Flush customers ordered big bottles. The cummerbund-wearing corps of waiters brought, opened, and poured but had no idea what they were serving and, when asked, knew even less about what to suggest. To help sales, the owner invested in midafternoon tasting seminars. He'd open some of the good stuff to taste, and the waiters would wade into the thicket of names and learn what those bottles contained.

For Wright it represented an awakening of sorts. German wine bottles—whose Gothic script and compound words had been indecipherable—now proved to hold a sweetness he'd never tasted before, as if suspended in a web of acidity. California became more than a distant sun-kissed place; the fruit pits and briars lurking in Ridge Vineyards' Lytton Springs zinfandel invoked the steep escarpments of some distant mountain. He even tasted rare and expensive first-growth Bordeaux. Nothing though had prepared him for the taste of Volnay Caillerets, a renowned vineyard from a small town on the acclaimed slope known as the Côte de Beaune. Often bottled directly by the producer, this one came from Bouchard Père & Fils, a *négociant,* or dealer, one of the renowned houses with broad wine holdings whose headquarters and large warehouses had made Beaune the logistical center of Burgundy. Where some of the other reds had a raspiness to them—tannin to an unaccustomed palate—this one had a certain liveliness. It was light but inescapably present, not concentrated yet powerful, as if every molecule was saturated with something wild or barely tamed.

Any lover of red Burgundy would recognize in Wright the incipient symptoms of obsession. Within a week he was in the Lexington Public Library, taking out books on wine. There were many thick tomes, but his favorite was Alexis Bespaloff's *The Signet Book of Wine*—a pocket-sized paperback he could read between classes or even sneak a few pages of while waiting for his shift to start. Reading Bespaloff's description of the narrow road—RN 74—that runs south from Dijon, skirting the slopes of all the great vineyards, Wright came to understand just how slim the demarcations were between villages and even between the vineyards of each commune. How come?

This was very far away from Lexington, Kentucky's second-largest city, with its generic brown-glass downtown office towers offering some kind of counterweight to the university's large campus. But the city was located in the middle of bluegrass country, and with legendary racing stables all around, the better liquor stores had enough of a carriage trade to actually stock some fine wines and at good prices. With Wright tracking down books and Holstein as an eager participant in the learning, the friends were soon saving up money to purchase the very wines they were reading about. Soon the two friends didn't just have opinions on Les Caillerets, but on adjoining vineyards such as Les Champans. The vines grew side by side, but the wines were deemed different. Each new word that entered their vocabulary got bandied about in their discussions. Did Holstein think the Beaune villages showed signs of Beaune's delicate *terroir*? Indeed, and would Wright not agree the Pommard displayed a certain characteristic *velours*? Back and forth it went for hours at a time. "We were shot in the ass with Burgundy," Holstein recalls of those tasting sessions.

Divided by village appellations and then subdivided by vineyard names and given a final layer of complexity by different winemakers harvesting individual rows of the same enclosure, Burgundy has the makings of a lifelong fascination. To bear a name, a plot of land must be deemed worthy of expressing something different from the adjoining equally storied parcel. The small village of Volnay has many legendary vineyards, each with its own name, each giving a nod to its particular setting. In addition to Les Caillerets, which alludes to small stones found in the loam, there's Clos des Chênes, which refers to a copse of oaks it once held, just as Les Brouillards evokes the mists known to form on that particular slope. These are age-old designations. In *French Rural History: An Essay on Its Basic Characteristics,* the French historian Marc Bloch argues that the *lieu-dit,* or place name, represented a necessary fixed area, useful for everything from giving directions to raising tithes to establishing census data. They might be "bounded by a visible limit," he writes, "a hump in the ground, a stream, a man-made embankment or hedge. But often the only feature distinguishing one from another was the orientation of their furrows." In Burgundy, it was vines not furrows, and in addition to helping someone know when to take a left, these specified and named demarcations established the idea of nuanced differences that the budding connoisseurs were only too eager to become familiar with.

Whether on the map or in the bottle, such a sense of rootedness was the very opposite of the life Wright had known. Though his parents had grown up miles from each other in the same Illinois town, by the age of six, Wright's father's job in marketing had repeatedly sent the young family on the road, always to the next job. As quickly as the products his father was marketing changed so did the towns. Walker mufflers, Mercury outboard engines, Rawlings baseball gloves . . . Fond du Lac, Racine, St. Louis—Wright had known them all. The one constant throughout the moves had been wrestling. He'd started in middle school and fought at a trim 126 pounds by high school. As a junior, he won the Illinois high school championship in his weight division and was primed for a great senior year, when he learned he'd be adding a name to the litany of companies and locations. Now it was the liquor giant Brown-Forman and Louisville.

Here he was now, a political science student at the University of Kentucky whose life had suddenly been taken over by the nuances of distant vineyards. As a graduate horticulture student, Holstein wanted to reconcile his interests in sciences and winemaking, so he proposed they grow their own grapes. Wright deemed the idea fabulous, and Holstein had soon talked the pomology department into allowing him to plant vines on a nearby vacant lot. The friends dug a trench and planted a row of Vidal, a white hybrid that they figured had a possibility of yielding fruit. Wright today laughs at the short-lived experiment. "Vidal may be able to take some humidity but not Kentucky humidity." What the experiment did make clear was that he wanted to study winemaking—before succumbing to black rot those vines had interested him far more than any course book. The idea that had slowly been developing in his mind was starting to take a very clear form: he wanted to be a winemaker.

He wrote to the University of California at Davis, which had the country's preeminent wine program, and enrolled in the fall semester. Late in the summer of 1975, he headed toward California in an old VW van listening to the Alan Parsons Project on a tape deck. His belongings consisted of a few of his favorite Nehru jackets and little else. It was hot on the drive. So hot that while motoring through Nevada, the boiler blew. It was 106 degrees when he pulled into Davis a week later—he noticed the reading displayed on a drugstore facade. The searing summer days made the Davis campus a natural place to study the role temperature played in winemaking. Investigating how a single vineyard might be unique, that was far more problematic.

UTTERED OFTEN IN WINE LOVERS' conversations, the name Davis suggests everything from academic rigor to hardheaded obstinacy. This leafy campus in Yolo County, west of Sacramento, is where generations of viticulturalists who plant vineyards and enologists who make wines have come to be trained. In a way, the academic environment at UC-Davis represents the counterpoint to the rapturous tone of modern winemaking that breaks down all those valleys, vineyards, and blocks of vines into uniquely expressive places. Pure sales talk is what that's called in Davis. In its simplest terms: the modern wine business draws a bead on a place to establish quality. Davis rejects the notion of a specific—usually tiny—locale being superior as scientifically unsound and unverifiable. It's remarkably easy to get at least an eye roll from either party: the academics barely suffer the talk of *terroir,* that French notion that implies uniqueness of flavor from delineated areas; the winemaker grows distraught that all those bristly headed scientists fail to understand the very lyricism of wine.

From the earliest days of agricultural research, the University of California was concerned with far more practical farming matters. Already in 1880, Eugene Hilgard, a professor of agricultural chemistry who led the young department of agriculture, was writing to Senator S. G. Nye in support of a viticulture bill the legislature would be voting on: "The growers need to know, and that quickly," he wrote, "which of the 2,500 varieties they should choose." For the German-born soil scientist—he received his doctorate at the University of Heidelberg under Robert Bunsen, of lab burner fame—the climatic conditions of California were the chief reason preventing the state's wines from reaching their rightful position. High heat meant high sugar and hard-to-drink high-alcohol wines. Drawing on the research capacity of the Berkeley campus, Hilgard quickly set about establishing principles to improve the wines. Soil drainage, winery practices, technical advancements (he was an early proponent of pasteurization)—these were all subjects studied, discussed, and put out to farmers through Agricultural Experiment Station bulletins. Alas, these were not always read in full—newspapers tended to publish whatever length they needed to fill empty space—but their spirit was practical and clear. In a 1914 circular on pure yeast strains that could function in the blazing temperatures of the Central Valley, Frederick Bioletti, one of Hilgard's first hires, writes as if he's giving directions personally to the winemaker. "Two flasks will be received from the laboratory, one marked 'Yeast' and the other 'Must.'" The article goes on to describe how to propagate this strain and thus complete fermentation even in early fall's intense heat. To the Fresno farmer with flies circling the redwood vats of a stuck fermentation, those flasks were the very tools that would get the whole process moving again. Practical information indeed.

Whatever advances had been made came to a screeching halt in 1920 after ratification of the Eighteenth Amendment to the Constitution prohibiting sale and manufacture of all alcoholic beverages. Today, when the speakeasy is the model for the modern mixologist-driven bar, the barmen in suspenders muddling herbs and rhapsodizing about small-batch, bonded whiskeys, it is hard to imagine what an enormous and detrimental effect it had. In Kentucky, warehouses were sealed; in Milwaukee and St. Louis, the breweries emptied vats, turned off the malting kilns, watched as government agents padlocked doors.

Almost immediately the wine industry turned to the only viable way it could legally continue by shipping hardy grapes to all communities with immigrant groups who could make their own. The process was simple enough. A freight train came to warehouses throughout the San Joaquin Valley—often a spur track even allowed it to trundle to the vineyards. There was a certain jostling by growers for the refrigerated cars, which would be topped off with ice at various points throughout their journey. Refrigerated or not, a stick of sulfur dioxide was burned to fumigate, and in went the grapes. Ernest Gallo, then selling his family's grapes in Chicago, paints a vivid scene on the tracks on Twenty-First and Archer, where jobbers "would buy the car and then open the doors, and sell it out in one hundred and two hundred case lots to their friends to take home and make wine."

Any advances the University of California had been able to make until then seemed utterly forgotten. William Vere Cruess, a zymologist, or expert on fermentation, who had previously furthered Bioletti's research into pure yeast strains, turned his attention to canned fruit cocktails. In Circular 30, published in 1929, nine years after the start of Prohibition, Bioletti strikes a rueful if unimpeachably scientific tone when he writes: "Formerly Cabernet, Semillon and Riesling received a considerable premium. Now they are almost valueless. The varieties that are the most profitable at present are those which have deep color and are sufficiently firm and tough-skinned to get to New York without spoiling. Viz. Alicante Bouschet, Carignane and Mission." Not a superior grape among them.

With the ratification of the Twenty-First Amendment came the repeal of Prohibition in 1933, and UC academics were once again free to instruct on their areas of expertise and throw themselves at Hilgard's great notion of what to plant where. To look back at that moment from the vantage point of today, an era when the microclimates are charted, the rootstock described, the wood of the barrels listed, the geographical and weather features expounded upon, it seems impossible to imagine what a neglected situation they were starting from. The vineyards were a mess of grape types with the coarser, shippable varieties predominating. The cooperage—all those unused barrels—was rife with *tourne*, a taste-altering disease. The chain of knowledge that linked generations had been broken in that thirteen-year lapse without a single vintage.

Transferring the viticultural department from Berkeley to Davis in 1935 brought a new measure of energy to the endeavor. And in one of those strokes of luck—which the wine industry had been in short supply of—two eminent scientists entered the department whose names and research are the stuff of lore. The two chiefly responsible for solving the what-grape-where question were a thorny Texan, the head of the department, Albert Julius Winkler, and a dimple-cheeked son of San Jose whose family had orchards in nearby Modesto, Maynard Amerine.

The two men could not have been more different. A native Texan who had never lost his turns of phrase, Winkler was known to ask for a steak "as rare as if it ran through barbed wire" at barbecues. More of an opera lover, Amerine was punctilious in his research—he'd taught himself French, Spanish, German, and Russian in order to read foreign research papers—and clear in his opinions. Over the next four decades, his initials, MAA, were to become the very stamp of authority on the post-Prohibition wine world, as they graced studies, introductions, and bibliographies, and when pushed, he became the source of withering quotations. "Ah, *terroir,*" he was given to say, "dirt is dirt."

Almost immediately the two men threw themselves into the research that would become the key study through which the

California wine industry would be revived. Since the growers often didn't know what they had in their fields—"They thought anything wrapped up in a grape skin would make wine," Winkler allows in an oral history—the two scientists would collect their own samples, ferment them at Davis, perform taste analysis of the untreated juice or musts, and, correlating the highest quality of these with the temperature they'd grown in, divide the state into regions that suited grape types best.

The growing infrastructure of California made these rapid grape-picking excursions possible. Winkler and Amerine prided themselves on gathering their five-pound boxes "from Ukiah to Escondido," essentially from the Oregon border to the Mexican one, and it's impossible to think of this occurring before the state's highways were built. In 1933, the Grapevine incline that links Northern and Southern California transformed miles of curvy roads that looped back and forth on themselves into one straight, smoothly graded ascent. The institutional knowledge that existed back in Davis also played an important role. When stumped by a grape type, Winkler and Amerine often turned to grape geneticist Harold Olmo, who would find paper bags containing seeds on his desk in the morning, something his colleagues left for him to decipher before starting off for other vineyards at dawn. The fermentation of all grapes were performed in a campus hut that had a hose perennially running over its tin roof to keep the temperature down.

Nothing was grand. And yet the two papers that resulted—1943's Circular 356 and the more academic version that appeared in *Hilgardia* in 1944—represent the cornerstone of the Davis approach to wine. "Composition and Quality of Must and Wines of California Grapes," as the more formal paper was named, compressed facts, analyzed reams of data, cataloged weather station readings, and divided California into five climate regions. The winemaker who dismissed the recommendations of what vines to plant where did so at his own risk. The range was massive. California's Region 1 equaled the Mosel Valley of Germany; Region 5 was the same as Algiers. If California wine was to improve, it would only be by getting the right vines into the right place to succeed.

In later years, Amerine would allow that, given time, they would have come up with more than the five regions, but for now the two papers represented a call to quality. The days when a farmer, hedging his bets until he saw whether it was more profitable to sell his harvest for eating or winemaking, could glibly call a table grape like Thompson Seedless the "Riesling of the San Joaquin Valley" were over. Whether it was by guiding, hectoring, or clarifying, the university and its faculty dragged Californian wine into an era of unprecedented quality.

FOR WRIGHT, AS A STUDENT, there was a constant awareness of knowledge there to be had. You showed up at a course—say that of bushy-mustache-sporting Ralph Kunkee—and there you were learning from the key professor in the study of malolactic fermentation. Sensory chemistry classes under Ann Noble gave him a first-rate understanding of the flavor components in wine. (He had a nose for amyl acetate—the flavor of banana.) Wright started lifelong friendships with future colleagues like Rollin Soles, a fellow student. In the evenings there were shifts at the Brewster House restaurant, where he introduced the postprandial *café à l'orange* and was kept busy making tableside Caesars.

Davis's academic standards were high and exacting, but Wright couldn't help feeling that the university existed slightly outside of changes that were going on in the broader wine world.

Mandatory reading, such as Winkler's *General Viticulture,* laid down principles completely foreign to his understanding of wine. "If the soil were so important a factor as is often claimed, there would be some uniformity in one or more characteristics between the soils and the various districts," Winkler writes. "The French viticulturalists, nevertheless, attribute great importance to the lime fraction of their soil. In the cool areas their opinion may be valid. The Italian viticulturists, in contrast, attach no special virtue to the lime content. In Germany, much benefit is attributed to the slate-stone and shale soils, even though many of their excellent wines are made from grapes grown in other types."

There's a barely suppressed pleasure in allowing the variety of opinions to undermine the very point they seek to prove. But there's also a creeping lack of relevance, the fate of anyone who, seeking to revoke a generally accepted principle, narrows a discipline to the single issue he obsessively cares about. Read then—and even more so now—such a view signaled being out of touch with the direction wine quality had taken. Joe Heitz, one of the winemakers who would come to represent California's new wine world, was no dilettante (he'd worked for Beaulieu Vineyard and taught enology at Fresno State University), but as early as 1966 he'd deemed the fruit he was growing in Oakville's Martha's Vineyard as exceptional enough to bottle under its own Heitz Cellar label. This is the great paradox of the Davis work: the upgrade in quality it achieved was enough to inspire winemakers to think of the viticultural landscape in new—or perhaps that should be *old*—ways. "We have harvested, fermented, aged out and bottled separately thirty different zinfandel vineyards from all over the state," Paul Draper describes in an oral history of Ridge Vineyards. "From Mendocino down to Paso Robles, over to the Sierra foothills, and everything in between." Where Amerine and Winkler had crisscrossed the state in order to create the foundation for quality, that role was now the winemakers', who sought, in picking from various parcels, to find individual expressions of the same grape.

It wasn't a transition; Davis was ceding nothing. But with the individual plot being promoted over vast swaths of land and grape's native yeasts increasingly being chosen over more flavor-determining pure strains, it was clear that a new spirit was at work. Perhaps some of these differences were generational. The scientists who brought California out of the post-Prohibition murk had barely enough time to get their bearings before the United States entered the war. "We gave twenty-six years to the armed services during World War II," Winkler says in *his* oral history, meaning twenty-six years of lost scientific research by men like Amerine, James Guymon, and John G. B. Castor, all of whom served in the army. It must have been fairly strange for the older academics to be in a wine world that was going at things in a very different way. In 1972, Josh Jensen, who would go on to found the Calera winery in 1975, was traipsing over the arid mountain peaks of San Benito County with a California State Mining Board geologic map and an eyedropper of sulfuric acid to see which rocks would fizz on contact with the liquid. When one did, the soil contained lime just like the great vineyards of Burgundy. That's where he wanted to plant his vines. Founded in 1976, Sonoma's Ravenswood winery had as its mirthful motto "No Wimpy Wines." In a photograph reproduced in Charles L. Sullivan's monumental *A Companion to California Wines,* a bearded Joel Peterson of Ravenswood is shown goofing around with colleagues celebrating the end of the vintage—each standing naked in a barrel. It could be the cover of a Richard Brautigan novel. At Davis, scientists shrugged. No hypothesis, nothing verifiable, and much talk about the individuality of certain hills—the new wine world.

PERHAPS BECAUSE WRIGHT TODAY IS so associated with Oregon winemaking, it is tempting to think of his time in California as nothing but a preamble. It is true that his arrival in 1986 in Oregon's Willamette Valley seems predestined. For a winemaker like Wright, whose fascination started with tasting the wine of a single vineyard, the layered compression of the Oregon landscape seemed a natural fit. In Oregon, history sits very close to the surface. Yamhill County, where Wright would eventually settle, derives its name from Yamhela, the Native American tribe that once lived on this land. One can cross a field waist-high in oats and come across the tombstones of settlers killed in the Cayuse War of the mid-nineteenth century. Roads are narrow—though not enough to slow down the loaded logging trucks coming down from the mountains. Smaller towns such as Yamhill or Carlton may have a single swaying traffic light. Wright's first winery was by the train tracks in McMinnville, the county seat. A disused Carnation egg plant, its walls were insulated, and, more important, it was a large enough space to fit a stemmer and stacked barrels. He called the winery Panther Creek, the name of a stream that jags northwest to southeast in an 1853 surveyor's map of the South Willamette meridian.

In California, it wasn't quite that easy to dip one's hand into the past. Still, Wright's decade as a winemaker there was more than a blip on the road to getting to Oregon. After leaving Davis at the end of his junior year, Wright soon got to see the versatility that a solid enological training gave him. His first job was at Ventana

Vineyards on the west side of the Salinas Valley, where at a certain stage he was making twenty-eight wines, ranging from chenin blanc to dessert Rieslings. His second job was not far away at Talbott Vineyards in the Carmel Valley, a small winery that specializes in chardonnay. While he was at Ventana, he'd been hired by Dick Graff, the legendary winemaker of Chalone Vineyard, to make its second label, Gavilan, named for the mountain range whose brown round peaks mark the eastern border of the Salinas Valley.

Here, in this corner of California, was an undeniable sense of place. In the Salinas Valley, where ramrod-straight roads cut through romaine and iceberg lettuce fields, or, in Carmel, where fog envelops wind-twisted cypresses, or, even more so, near Chalone, east of the prison town of Soledad, where the dry afternoon air carries the sage, chaparral, and coast live oak it has crossed, there is a quality to the very air of each locale that is indelible and unique. Wright, who, paradoxically, rarely uses the word *terroir*, traces his quest for an American version of that uniqueness to those days. He was becoming a winemaker, proficient and confident. But the wines that he responded most to had a deeper quality than simply reflecting the landscape. There was a rooted sense to them, as if they'd issued from somewhere with firm foundations. That was possible, even in modern California. Wente, Martini, Georges de la Tour, Mirassou, Gallo, Gundlach Bundschu, Rossi, Ficklin—these were names close enough to the foundations of the state's wine industry to be founders. Cesare Mondavi had come to the state during Prohibition to buy grapes for an Italian social club in Minnesota that needed regular shipments of fruit; his son, Robert, had been so taken with the architecture of *Sunset* magazine's headquarters in Menlo Park he'd asked the architect, Cliff May, to design a building for the Robert Mondavi winery with a sweeping iconic arch through which people would enter to sip cab and listen to jazz. The Roaring Twenties to the Blissed-Out Seventies, that too was an arc.

Were such transformations even possible anymore in California? Wright wasn't coming from a different industry, his wealth made. He was a father and a husband but also a hired hand—one who was ready for the move to Oregon when the moment came. Winemaking there was moving in the direction of the fine French varietals, but there was still a far more folksy tradition of wines such as the loganberry and rhubarb wine Henry Endres had been making up near the Clackamas River for decades. They were local; they might even have some of that *terroir.* HONK FOR SERVICE read the sign welcoming those who drove up for gallon or half-gallon jugs. When Wright's old friends Allen Holstein and Rollin Soles (Soles had earned a master's in enology and viticulture from Davis before working in Switzerland and Australia) moved to the town of Dundee to help launch and manage the Argyle Vineyards, it felt like destiny.

By then winemaking was established enough that it could be divided into eras. The title of pioneers went to Richard Sommer, David Lett, Dick Erath, Dick Ponzi, Susan Sokol Blosser, and David Adelsheim, who, from the mid-'60s to the late '70s, had taken augers to the hillsides to plant vines, driven old Massey Ferguson tractors between the rows spraying them, thinned the grape clusters to concentrate the flavor, and transformed the compressed power of those grapes into wines that were increasingly becoming recognized. Though his reputation as a winemaker never reached theirs, the individual most responsible for there even being a wine scene was Charles Coury, whose 1964 master's thesis at Davis had established Oregon as a new frontier for wine and, in particular, pinot noir.

Using the theories of U.S. climatologists who after the war had to decide where to plant wheat in a hungry Europe with the certainty it would grow, Coury set out by asking how European grape varieties became indelibly associated with certain areas. The answer for the original choice of varietal was clear enough, "the result of generations, even millennia of patient selection and refinement by countless, canny European peasants." But there was a twist. In each area Coury studied, the variety chosen ripened extremely late. What was that about? Surely a canny peasant wanted a grape he could have in the hopper before autumn winds threatened. "Everywhere in Europe the same pattern repeats," he writes. "Varieties have been selected which in the average year just barely attain maturity." The only answer he could come up with was that at the precipitous moment when all might be lost, the absolute of quality could be achieved.

Coury hadn't been thinking of Oregon in the dissertation but rather how European principles could be applied to California. The figures he gathered were all from microclimates in Napa and Sonoma. In the notes, Amerine—who else?—is thanked for opening up the Davis cellar for the necessary taste tests. But the theory that grapes were at their most expressive when most endangered had been enough for adventurous winemakers to found a new region. Fully eight degrees of latitude north of Napa, the Willamette Valley supplied all the climatological danger a winemaker could want. Come fall there is a moment when the storms are backed up toward the Aleutian Islands and the grapes are still on the vine and the harvest will either be great or a watered-down loss. If you were ready to risk it all every year, you could make great wines.

Wright first came to visit in 1982. Seeing Holstein driving the narrow roads where hazelnut and plum orchards ran up alongside vineyards, he felt grounded in a new way, as if this was a landscape with potential for him. Place is a curious thing. In the essay "Some Notes on River Country," Eudora Welty reminds us how "a place that ever was lived in is like a fire that never goes out. It flares up, it smolders for a time, it is fanned or smothered by circumstance, but its being is intact, forever fluttering within it, the result of some original ignition." That was easy to feel here. There were no châteaus or stately stone wineries, but a gambrel-roofed barn had its own heritage. Even the red coat of paint on the barns that dotted the hills is unique to the American experience since the color was originally made from linseed oil and red oxide of iron, two ingredients any farmer could get. That butte that served as a marker in the distance? It's named after Alec Carson, a relative of Kit Carson, the celebrated frontiersman.

On the day Wright and Holstein sneaked into the Erath winery and tasted the '83 and '84 from the barrel, the circle seemed complete. Here was that featherlight intensity they'd tasted as students. Wright wasn't thinking of Les Caillerets or Clos des Chênes now but of all the places he'd lived from St. Louis to Fond du Lac to Salinas, as if those places were way stations en route to this goal. All those years of wrestling had taught him to deal with a huge variety of opponents. Often he'd never even have seen whom he was to wrestle before stepping on the mat. "It might be a tall gangly kid or one built like a V," he recalls. He knew he was ready for this new challenge. Less than a year later he was driving his family up in a truck from California, their belongings following them on Interstate 5 in an Allied van.

TO MARKET

Paul Saginaw and Ari Weinzweig had only one requirement for the deli they opened in March 1982 in Ann Arbor, Michigan. All the food had to be excellent, or it wasn't worth doing. Zingerman's Delicatessen was thirteen hundred square feet of a strangely shaped corner building situated in a tangle of streets still paved with cobblestones from when the neighborhood had housed the city's draft-horse stables. By 1982, Kerrytown, as the old Fourth Ward was known, was home to a flophouse, a pool hall, and a couple of erudite bookstores; the one stab at quality was a fish market that had recently opened in a reconverted old redbrick structure on the neighborhood's edge. For gastronomy, this was unknown territory, but so was the rest of the city. Ann Arbor may have been home to the prestigious University of Michigan, but at faculty parties, the wrapped sausage pig-in-a-blanket was still served as a gourmet treat.

Saginaw and Weinzweig had met at one of Ann Arbor's recent efforts to raise the quality. They were both employees of Maude's, a restaurant with a belle epoque theme across from the recently opened bicentennial post office and transit center four blocks from the University of Michigan campus. The restaurant had been started by Dennis Serras, a native of Schenectady, New York, who had drifted to Ann Arbor in the late '60s on a student deferment. After a couple of odd jobs he went into the restaurant business in 1975, launching the Real Seafood Company on Main Street. In 1977, there was little that was keeping downtown active. Certainly, there were the businesses that might take a freshman from his first pair of Hush Puppies as an undergraduate to the basic suit he'd need in the workplace (and there were always businesses that sold pens, trophies, and boxed book sets to visiting parents), but like so many other American cities in the mid-'70s, Ann Arbor was losing out to its suburbs.

Serras hired Paul Saginaw as manager of Real Seafood, seeing in the Detroit native the kind of Runyonesque affability that would leave Serras free to pursue other restaurant ventures. In fact, two years later when he took over the lease of the Golden Falcon, a rough saloon around the corner on Fourth Avenue, and transformed it into Maude's, he transferred Paul there to help with the launch. The venture was a bit of a reach. Maude was ostensibly a madam in a turn-of-the-century bordello—her Aubrey Beardsley–like portrait graced the menu—and salads such as the Juliana, Diana, and Hortense were the names of her girls. Otherwise, the feel was relatively bistroish. The waitstaff wore black pants, white shirts, and black aprons. All dinners came with Kaiser rolls; the signature dessert was an amaretto mousse. Though the prices were relatively modest—$8.95 for the king crab legs—they were too expensive for the majority of students. It wasn't exactly what Serras had intended, but the long happy hour provided a haven for postal workers who trailed over from across the street when their shifts ended.

A TEAM SPIRIT DEVELOPED ALMOST immediately among the staff at Maude's. The hours were long—the bar remained opened until midnight. As manager, Paul, resplendent in a white poly-blend suit, might perform a pirouette to Barry Manilow's "Copacabana" as he led a party to a table, just to amuse the cooks peering from the kitchen's pass-through window. Frank Carollo, a recent graduate of the University of Michigan's engineering school, worked as a cook and wielded a fierce bat on Maude's softball team. He was helped by Ari Weinzweig, a rail-thin Chicagoan with a mop of hair and a history degree who had entered the restaurant business as a dishwasher while he figured out what to do with his life. The future and all its concerns disappeared behind the gauzy veil thrown up by the pleasures that the present held. Time didn't press. There was work, and there was skinny-dipping in the nearby Huron River afterward. Nineteen seventy-eight was the summer of Eric Clapton's *Slowhand* album, which blasted from the city's boom boxes as the Maude's crew gelled. When Maggie Bayless was hired as a cocktail waitress, the nucleus that would go on to found Zingerman's four blocks away and four years later was mostly complete.

Weinzweig had applied several times before getting the job at Maude's. A friend of his was one of the revolving cast of waiters and Weinzweig had thought he'd join in. Turned down as a dishwasher, he tried again a few months later and was accepted for the job. Like the majority of the staff, he felt that he was merely treading water. All had degrees—some advanced degrees—and all were waiting for the next chapter in life to begin. Weinzweig knew he didn't want to return to Chicago where he'd grown up in a kosher-observant home. He might even say that in the hard toil of dishwashing he'd found an extension of his studies. While working toward a degree in history, he'd specialized in Russian history with a focus on anarchists, a field of study—and soon a fascination—that had led him to the recently opened seventh-floor annex of the Hatcher Graduate Library, where the university kept the Joseph A. Labadie Collection of anarchist texts. Seated at the long reading tables that indented the forest-green carpet, he found a place to while away the hours.

The University of Michigan had a well-established tradition of political unrest. The term *teach-in* was coined in March 1965 to describe an all-night information-gathering on the Vietnam War in one of the buildings overlooking the central quad. In the subsequent years, the legendary Diag, named for the diagonal paths that led between departments, became a blank slate where students might demonstrate against the ROTC's presence on campus, dig a crater to create an allusion to the American bombs that were being dropped in Southeast Asia, or gather before marching beyond the campus border for the inevitable showdown with the Washtenaw County Sheriff's Department. In his reading at the Hatcher, Weinzweig was absorbed by an earlier struggle. He read a first edition of Emma Goldman's autobiography—with its transporting handwritten inscription: "To Fabionovitch who was a baker"—and the seminal *Fields, Factories and Workshops* in which Peter Kropotkin describes a "mutual aid" society founded on the ideal of greater good. Tome after tome each held something different and significant. Weinzweig had a relationship with the *guichet* window the books were passed through like a hummingbird has with a syrup dispenser—constantly requesting books as he read his way through the anarchist canon.

There was something in Ari's work ethic that would have pleased those early anarchists. Four years after starting at Maude's as a dishwasher and then being promoted to line cook and manager, he became manager at Serras's latest venture, a restaurant called Mantels ("Eight rooms, eight mantels" was the no-nonsense theme) by the newly constructed Briarwood Hilton, out near Ann Arbor's small airport. The little Maude's group had broken up. Frank was at Real Seafood, Paul had left to open Monahan's Fish Market in Kerrytown with Mike Monahan. When the Market Place deli across the street closed, Paul contacted Ari and brought up an old idea. Maybe they should open the kind of Jewish deli they'd grown up with but have it go beyond the Hebrew National salami in the cold case, the good thick sandwiches, and the craggy block of halva. It was an idea to consider—ideally over beers at the Old Town Tavern when the group reassembled after their shifts at the various restaurants.

The scene is worth pausing at. After all, it was the moment they joined the real world—those heroes of nineteenth-century Russian literature seemed to constantly be at that precipice. The student who has become the ironic ex-student finally chooses a path of action. It was easy to laugh about restaurants with mantelpieces for a theme, harder to leave them. When they took over the lease of the deli, they threw themselves into the renovations with enthusiasm. They kept the old Formica tables, and Paul bought a much-used classroom blackboard from the university's material department to write the wry sandwich names on. In recalling their early days, there is a measure of pride in the lack of airs. "We took food stamps," says Ari, "and sold milk and cigarettes." A chipped and endlessly repainted bench sat on Detroit Street right by the entrance; as far as anyone knew it dated to 1902, when the brick building's first incarnation had been Disderide's Grocery. The name they made up—Zingerman's—sounded right, too.

The low-key approach fit into the city's ethos. Customers could sit, eat, converse, or daydream with a backdrop of dry salamis, which hung from a pipe behind the cheese counter. It was in this mix—one customer at a table reading the box scores of the *Detroit News,* another poring over William Empson's *Seven Types of Ambiguity*—that Zingerman's took shape. There was a direct relationship between Saginaw and Weinzweig's own learning curve and the foods they offered. Ari, who had soon established himself as the purveyor, would find something that excited him and soon it would be on the blackboard or somehow stacked in the impossibly tight space between aisles. At Zingerman's the staff didn't believe foods had to come from Europe to be good. Nor did American gourmet mean Californian. Corned beef from Detroit's Sy Ginsberg went into the "Who's Greenberg Anyway?" sandwich; thick, crisp slices of Wisconsin's Nueske's applewood smoked bacon were piled in the BLT on rye.

Yes, the Jewish deli was the template for Ari and Paul, but it was a model to be expanded upon so that Red Pelican mustard, Carr's crackers, and Walker's Scottish shortbread could also be stocked. For good rye and challah they went to Modern Bakery, forty miles away in Oak Park. (Maggie did those runs, driving in the predawn hours in an old van with a broken fuel gauge.) Given the exacting nature of the modern artisanal selection, there was something extremely accessible about what they offered. There was nothing too refined about it: the blue cheese was Saga; the olive oil, Pope or Colavita, they ordered from the old Italian delis in Detroit. At Zingerman's, quality took many forms—some unexpected. When Paul wanted to run egg rolls on the menu, he'd buy them from his friend Wai-Chong "Johnny" Tam, whose Chinese restaurant, Middle Kingdom, stood across from Real Seafood on Main Street.

If there was a unifying principle to the selection, it was that these were fairly priced, fully flavored foods. This concept seems more obvious to us than it did decades ago when the timid flavors of the American table evoked all manner of condescension. One thinks back to the protestations of a Gervas Huxley who, in his 1956 tome, *Talking of Tea,* describes an Englishman receiving a cup of tea on our shores: "Confronted by a cup of rapidly cooling water—which was never boiling even when the cup was filled in the kitchen—and, in the saucer, by a tea bag on a string, which the drinker is supposed to dangle in the cup until the tepid water receives a brownish tinge . . ." But Zingerman's was doing more than providing a heightened sensory experience, it was rebelling against the predominant model of the grocery store. Though it is rich with folksy associations—"Advertisin' don't take the place of dustin'," one old-timer opined in a turn-of-the-century trade journal—the American grocery store was the end point for well-established chains of distribution. Shelf space was a realm of conflict almost as soon as there were large enough companies to stock them. Manufacturers were pitted against wholesalers who often created their own in-house brands; individual grocers (endangered already in the 1890s) depended on industry to underwrite their advertising, print their flyers, and even provide the paper bags—stamped, of course, with the names of their own goods.

When Zingerman's started, it wasn't like there was a parallel provisioning chain. They couldn't opt out of the system that existed and stock their shelves with teas, olive oils, and cheeses by some other means. Finding quality ingredients represented their earliest important challenge. At first, Ari had thought they could use the food distribution system that existed to expand their selection of foods. When he called, though, most wholesalers told him the orders were too small to bother even sending a truck to Ann Arbor. And, no, they weren't going to track down a cheese for him. They had their lines of product and that was what they sold. "I've got as much interest in carrying that cheese," said one distributor, "as the man in the moon."

A second difficulty was getting the clientele excited about some of the products they were starting to carry. Education was the key, Ari believed. They had to establish the idea of quality being something that is worth paying for. And to do that they had to learn more. Much in the way Ari had once looked at the anarchist tomes in the Labadie Collection as the reservoir of truth, he now set about reading food articles from every publication he could lay his hands on for information. There was a bigger food world than the one controlled by distributors. Perhaps people thought they didn't need what Zingerman's sold, but that only meant he and Paul had to find what people wouldn't be able to live without. His reading was prodigious. In what passes for the Zingerman's archives, pallets' worth of papers in a warehouse near Ann Arbor's small airport, one comes upon files and folders of clippings and torn-out articles from *Gourmet,* the *New York Times Magazine,* the *Detroit Free Press,* and the *Plain Dealer* in Cleveland. Weathered and brown, the food section pages hold recipes for quiche, ideas for sandwiches, varieties of mayonnaise, and anything on cheese, a particular interest of Ari's.

He'd read about Comté, the semihard cow's milk cheese

from the Jura Mountains along the French-Swiss border, and eagerly ordered it. What the truck delivered was something that had sat in a warehouse for months, slimy from condensation. Where were the incredible nuances of milk, pasture, and gradations of aging this cheese was capable of? He'd find small-batch cheeses—but Zingerman's couldn't get them delivered. It made him try all the harder. Ari could accept that they might be perceived as a weird-sounding, counterculturally inclined, college-town sandwich shop, but weren't they trying to tap into a real and traditional layer of American agriculture? Ig Vella had been making his phenomenally nutty Jack cheese in Sonoma since 1931. The Maytag family had been making a sharp, almost briny blue in the fields of Iowa since 1941. Farmers were organizing into cooperatives in Vermont, Oregon, and Wisconsin at Cabot, Tillamook, and Land O'Lakes, respectively, allowing their common interests (and when he thought of it, their "mutual aid") to build a business that allowed farmers to stay on the land. In the mid-1960s, the Wyndham Foundation had resurrected the cheesemaking traditions in Grafton, Vermont. And now, there were all sorts of newly minted enthusiasts like Miles and Lillian Cahn, who started Coach Farm with a herd of goats and a barn in 1985. Why was it so laborious to get these and so many other cheeses into the cheese case? Ari was not going to simply accept what the distributors thought was good enough.

It is perhaps useful to return to an earlier era of the grocery business to understand precisely what the Zingerman's crew was attempting. In one of his many paeans to the American country store, Gerald Carson, the poet laureate of the form, grows lyrical, describing their uniform scent as a combination of "sauerkraut, sweet pickles, the smell of bright paint on a new toy, kerosene, lard, molasses, old onions, potatoes, poultry feed, gun oil, rubber boots, calico and dried fish." Here, "the great wheel of Herkimer County cheese sat under its wire screen cage" amid a clutter of fly traps, peanut roasters, and coffee grinders and "a suspicion of corn whiskey rose from among the gentry sitting stove-side."

But there was a secondary role for these stores in addition to creating a gathering place, community message center, parcel room, and, in many cases, serving as a U.S. post office—they introduced people to new goods. In a fascinating study on a single grocery store in Pennsylvania, Diane E. Wagner refers to the role of the grocery store owner as a "culture broker." Once that role involved offering Chase & Sanborn coffee, Quaker Oats oatmeal, Eagle Brand condensed milk to customers unused to, and perhaps suspicious of, foods that came in sealed containers, but by the early 1980s in Ann Arbor, those world-enlarging foods included Michigan's own American Spoon preserves, Maille mustard, or a tangy Greek feta. The way Ari searched for quality harkened back to a time when the grocery business allowed for excitement and discovery.

In the fall of 1984, an employee arrived who would help them tell that story. A fine arts graduate of Michigan State University in East Lansing, Steve Muno had come to Ann Arbor to follow his girlfriend (and future wife), Monique. She was in her final year in college and he had a year to fill; he got a job at the scrappy little deli. At the time, anything that needed to be drawn or written for a sign was usually done by Paul's wife, Lori. Though Muno had studied photography and printmaking in East Lansing, he'd grown up with handwritten print. His mother had a home calligraphy business specializing in producing elegant wedding invitations and yearbook inscriptions for high schools in and around

Detroit. As soon as he was a teenager he was helping her do some of the more taxing yearbook stuff like cursive writing following the arc under oval portraits. Studying fine arts had only increased that skill, and it didn't take long for the Zingerman's staff to discover this talent in their new counterman.

Almost from the start, Ari had instituted a system of having the staff write little informative cards that were pinned onto a cheese or taped to a shelf right under the goods. The level of information was simple: a name, a place, perhaps what you could do with it or what it went well with. The tone was light, details were key, and Muno was good at it. His style was loose but clear, it generally avoided the uppercase, and it didn't really keep to straight lines; there was something about it that was jaunty and accessible. His notes on food took any possible stuffiness out of gastronomic information and made the cards seem more like tips a friend was passing on to you. It wasn't long before he was doing the signs for the entire store and before someone had christened the style Muno Bold.

Is it as simple as handwriting? Much thought has gone into the meaning of fonts and the psychological signals they send. Muno hadn't just grown up writing up yearbooks but had studied Soviet Modernist posters in college and collected the cover art of albums by the Smiths, the Clash, the Specials, and XTC—each one representing an example to be followed. And he knew enough about the finished look of the commercial art business to actively define himself against it. When he'd first decided to go into arts, he had studied advertising and it seemed entirely focused on winning people over by whatever means possible. "The only class I flunked in school," he says, "was a graphic design class."

In addition to the informative cards he was soon writing, he started turning his attention to the deli menu. From the very first days, these menus had been a single page listing the constantly renewed list of sandwiches. But with the amount of information that Ari was giving Steve to incorporate into boxes, and entire paragraphs that went alongside goods Ari was particularly excited by, the menu had to grow. Muno contacted a printer in East Lansing who did a lot of student union printing and requested a mock-up of a prototype publication Muno had designed. The back page was where the deli menu went, an inside page carried the full sandwich selection, and when the newsprint broadside was opened up to its full double spread, it recounted all manner of news and anecdotes about that month's products. The initial printing run wasn't great, but it was enough to stack by the door, to give every person who was waiting in line a copy.

The newsletter-menu also turned out to be just the thing to promote themselves as soon as they started attending the Fancy Food Show in 1984. Steve, Ari, and Paul visited this twice-yearly gathering as complete beginners. Even in the bright-colored '80s narrow ties they donned, they didn't look the part among the nation's better grocers all searching for new goods. At the show, there was talk of building your business through mail-order catalogs—Zingerman's launched theirs in 1993. They attended panels on employee management. Paul still remembers the answer Russ Vernon, owner of West Point Market of Akron, Ohio, gave at one of those panels. Asked whether he worried that he'd take the time to train employees about the foods and then they'd leave, the old-time grocer answered, "No, I worry I won't train them and they'll stay."

Armed with the newsletter, the Zingerman's entrepreneurs

were able to convince distributors that they were far bigger than they were. Here were talks, floor displays, receptions, and tastings going on in this tiny little Ann Arbor deli. What sort of distributor wouldn't want to participate? With each manufacturer it was important that Ari and Paul wrangle enough free goods that they could offer generous samples at the store. It was a new business model based on the idea of getting people to taste. "Ari was real good about just engaging suppliers," recalls Muno. "He'd say, 'Come on and do a tasting,' or 'Come and teach my staff,' 'Tell us what makes it special.' So we'd make an event of it, we'd do talks in the middle of the shop. I'd do press releases about anything we had going on. At one stage we realized we were going through five hundred pounds of corned beef a week; that was a press release. I'd mail it to the *Ann Arbor News* or WUOM and sometimes it was picked up. Stuff happens when you feed people free food, so sometimes we would have a couple of hundred people next to the building where we'd have a barbecue going making a paella, folks watching something exciting happen and learning all at the same time."

Here the newsletter worked wonderfully. A person reading about tastings, classes, talks, and presentations could have thought they were in the Harrods food hall of the Midwest. The great English cheddar producers sent truckles, the traditional clothbound shapes that the Zingerman's staff piled and presented—Muno did the Union Jack posters. (A picture shows Ari smiling from the middle of the display.) The Greek olive oil council must have been happy when the whole back wall was covered in butcher paper and painted with a mural of a white village against a turquoise sea and sky. It all sounds like fun, but something else was happening at Zingerman's. They were learning a look and language that went with their approach. There had been a set standard to what gastronomy conveyed to customers. The heavy paper stock, the glossy photos, the language heavy with foreign names, the overall idea being that what was imported was always best. Zingerman's took the opposite tack. Muno's first sign was an example of that. It read GOAT CHEESE FROM SHELBY, MICHIGAN, and was illustrated with a barn, two silos, and a goat that looked like Pokey ("The farmer looked like Gumby," Muno recalls). Underneath it read, "Uniquely mild and creamy, great for salads and sandwiches or just eating. A wonderful example of cheesemaking at its best. Ask for a taste." The tone here wasn't distancing; there were suggestions for how to use each food and an invitation to request a sample.

The style didn't just have a font. It had a cadence; it was warm, lacked pretension, delighted in detail. To read even the very first newsletters is to realize that this style was like a new language—one around which entire ways of thinking about food, and selling it, would form. Those early newsletters with their sparsely chosen adjectives, the importance of place names, the telling detail, the emphasis on small scale are like the Linear A of modern artisanal food description: the place, the helpful hint, the low-key user-friendly dish it could be put to. It was a style that evoked the small and unique and would become embraced, copied, and marketed by an entire artisanal food industry.

A new language of sales was being born, the culture broker idea given new life in a different era. Any food deemed worthy of being in the cold case at Zingerman's was capable of producing a historical step back: what cheddar was to the Pilgrims, or

Swiss cheese to early Wisconsin. Here, balancing the news of some recent discovery, were shout-outs to storied producers such as Mossholder, Ig Vella, or Maytag so that great food became something continuous, not a realm divided into old and new. All descriptions now contained the full Zingermanesque lilt, as laden with verbs as adjectives and with a telling, quality-related detail taking center stage: Shelburne Farms cheddar is from the "raw milk of the farm's herd of Brown Swiss cows"; Coach Farm chèvre is special because the curds are "hand ladled into molds (as opposed to being extruded as is done in most large goat cheese producers)"; at Brier Run Farm the flavors change "as the pasturage and weather change." If that cadence sounds familiar, it is a reflection of just how widespread it has become.

Zingerman's was applying it to vinegars, teas, coffee, rice, pastas, and jams by 1986 when Paul and Ari decided they had to expand the original store with a seven-hundred-square-foot addition to display ever more goods. The next expansion, in 1991, was to handle the long lines that gathered on the cobbled streets and wound their way around the block. Sure, people inching toward the sandwich counter represented an opportunity to turn them on to Andalucian olive oil or Portuguese sardines, but it was an inefficient use of space. By transforming the clapboard building next door into an annex, the finished sandwich could be brought to them in a different area—one where there'd be plenty of room for new goods and samples. With the start of the catalog in 1993, Muno's tasks that had begun with him writing index cards and making posters in his bedroom had transformed into overseeing a five-person art department in a separate building on Fourth Avenue. Gone were the days of buying art supplies from the student bookstore; gone, too, were those months in 1984 when he'd first seen a floppy disk and he taught himself to do graphics on the Macintosh Paul had brought in. His font—friendly, unpretentious, inviting—had framed an identity. Now it was all about broadcasting it. There were many more people than those waiting in line hungry for Zingerman's products and news.

TO TABLE

Atlanta was riding high in 1978 when a group of French chefs descended on it for a night. Having Jimmy Carter, the former governor of Georgia, in the White House created an overarching impression of power, but the city itself, constantly being built up and spreading outward, very much represented the latest version of the New South. Nowhere was this vitality more apparent than among the structures that had gone up in the swath left after the famed Beaux Arts–style Terminal Station and its adjoining yards were torn down in 1972. In close proximity—and linked with enclosed walkways for those wishing to avoid the draining heat—stood the arena of the Atlanta Hawks basketball team, the convention center, multiple parking structures, and the Omni Hotel. It was in the Omni's second-floor restaurant above the skating rink that the chefs set up for the all-expenses-paid, soft-drink promotion banquet they had been brought to Atlanta to cook.

They were a group of four and, despite being one more than the Dumas originals, liked to think of themselves as musketeers.

All were from the Gers region of southwest France; all were masterfully capable of transforming the regional bounty of truffles, Armagnac, foie gras, and confit into memorable meals. André Daguin, the proprietor of the Hôtel de France in Auch was their organizer and leader. Playing Porthos to Daguin's Athos—to continue the theme—was Maurice Coscuella, chef and owner of Ripa-Alta, a pleasant restaurant located by the colonnaded main square of Plaisance. The third member of the group was Roger "Zizou" Duffour, a specialist in roasting tiny ortolans, whose restaurant Le Relais de l'Armagnac stood in the small town of Luppé-Violles, a hamlet located in open country where the rolling hills of the Gers start to become the foothills of the Pyrenees. The youngest member of the group was Jean-Louis Palladin, a thirty-two-year-old chef who had managed to gain two Michelin stars at La Table des Cordeliers in Condom, a medieval town on the banks of a tributary of the Garonne.

In later years, Jean-Louis—that is what he would be called by all who knew him—would quip that the town's name was easy to remember. That was a decade in the future when journalists would be following him around, writing down every inconsequential thing he said. By then, his Washington restaurant, Jean-Louis at the Watergate, would be celebrated for bringing a never-before-experienced level of quality to the nation's capital and for drawing on American ingredients to create an entirely new kind of cuisine. His cooking wasn't purely French; it was too unencumbered. It was a paradoxical thing of reaching a French ideal through the best American ingredients.

That wasn't always easy—an early profile writer pictures him staring in shock at the tired button mushrooms found in a Georgetown grocer—but soon he'd found the quality that was available. "Sourcing" for Jean-Louis was a contact sport, and the tales of how he went about it amount to a gustatory montage. There's Jean-Louis slamming to a halt by a Maryland farm stand to load up on Silver Queen corn. And here he is at Arrow Live Poultry in downtown D.C. getting his chickens plucked or at a Chinatown fishmonger wrapping a live conger eel in a towel to turn into a Bordelaise specialty: the red wine stew called *matelote*. When he spotted someone who had reached a modicum of quality, he pushed them to excel. The forager who came in with a Budweiser carton full of assorted gilled caps was told they might have a deal if he could get more hen-of-the-woods mushrooms. The farmer who'd gotten into squab raising was instructed that the birds Jean-Louis received couldn't be more than twenty-eight days old. In Maine, there was a man in scuba gear walking the ocean floor and hand-harvesting the unblemished scallops Jean-Louis wanted. Only a retired State Department officer who had taken up herb gardening received immediate praise. "I just want someone to appreciate them," Richard Ober said at the kitchen door as he showed Jean-Louis baskets of chervil, sorrel, and the thinnest chives. "Did God send you to me?" answered Jean-Louis when he saw the sprigs. "You are exactly what I've been looking for."

WHAT BROUGHT THE CHEFS TO Atlanta was promotion at its craziest. The soft-drink division of General Foods had recently launched a canned lemonade called Country Time. This was not Kool-Aid, which General Foods sold in packages, nor was it the powdered Tang, which had the connotations of a recess snack. Country Time was intended to be different. Bill Burgess, the General Foods manager for the soft-drink business unit, was heading the campaign, and he sought to position it as something classy.

"We wanted it to be what you drank on the eighteenth green," he recalls. "The drink that wouldn't give you a carbonated belch."

Country Time was not Burgess's first campaign, and he knew that its success depended mainly on what the bottlers thought of it. Utterly dependent on huge companies and yet curiously autonomous, these bottlers were not going to clog up their production schedules with a dud. General Foods was providing them with nothing but the concentrate. The bottlers provided the water and the fizz to make it a beverage and the transportation to the stores. Without the budget of Pepsi or Coke (who would view Country Time as a competitor to its Minute Maid brand), he had to do something that would get the bottlers' attention. Their convention was being held in Atlanta. General Foods was going to invite them to a party they wouldn't soon forget.

CHEFS TODAY ARE CONSTANTLY BEING faced with proposals for promotions, festivals, opportunities to push brands while burnishing their own. But this 1978 event glimmers with something unique; it was perhaps a precursor to the kind of chef endorsement to come. Here, the chef—the French chef, to be precise—was an exotic object whose very presence could back up the wildest claims. After all, who better to advance the gastronomic claims of an upwardly mobile syrup than a band of Gascon chefs?

In the decade Burgess had worked at General Foods he had become something of a specialist in "fighting brands," brands that went up against far greater adversaries—and required unorthodox steps. A competitive spirit was what drew him to the food conglomerate to work after graduating from Pennsylvania State University in 1965 with a bachelor's degree in economics and while pursuing an MBA in marketing. He'd interviewed for jobs and gotten offers at Dow Chemical, IBM, Ford, Hallmark, and General Foods. He took the General Foods offer because it was the one where the competition was closest to the surface.

Shortly after arriving at the corporate headquarters in White Plains, New York, a brick structure with a gleaming white marble facade, Burgess found out just how competitive that was. The rivalry was fierce enough inside the building where product managers—MBAs all—jealously guarded their brands while keeping an eye on the success of others. Hearing Grape-Nuts lauded for their quarterly figures was not great for the Jell-O or Minute Rice folks. It was not all for one and one for all. As for fighting over the nation's palate with other companies, that was all-out war. The country was divided into tiny precincts, whose shopping habits were tracked down to the zip code by Nielsen ratings and corporate research. That sounds slightly vague, like something occurring in the realm of cool analysis. It was anything but. "We had ex-FBI employees on staff," Burgess recalls. "If Nestlé was giving out coupons to Taster's Choice in a particular zip code, we put three-pound samples of Maxim freeze-dried crystals on every doorstep so they wouldn't use them." The goal wasn't to affect the breakfast habits of this particular neighborhood, it was that this particular neighborhood was the smallest unit of the bigger battle for shelf space, ratings, buying habits, and market share.

With Country Time, Burgess was worried. He had been in the beverage division long enough to know how it worked when you had a budget. Yuban, sure, you could give away samples. For Maxwell House, General Foods held nothing back. A favorite ploy was to buy what was called a roadblock, an ad that was broadcast at the exact same time on all three networks. Burgess had watched those commercials for Max-Pax, a ground coffee filter ring from the Maxwell House division, on monitors in suites at the Americana Hotel in New York City where General Foods brass celebrated the moment. He had a $110,000 budget for the Atlanta event to promote Country Time. It was minuscule. But he didn't have to create a miracle; he just had to get the bottlers excited enough about the product they could foresee it having 1 percent market share, the figure that would get Country Time onto their trucks.

The plan of action had been hatched in a realm of food promotion beyond Burgess's experience. When first contacted about Country Time by the General Foods consultant, Henry Stampleman, a storied PR professional (who also worked for Coke and Pepsi occasionally), had understood that this was going to require something classy. He'd called Roger Yaseen, the American head of the Chaîne des Rôtisseurs, one of several dining societies popular at the time, to get a culinary lead. A trim Wall Street professional, Yaseen had no illusions about how promotions worked. "The Chaîne operated as a clearinghouse for these kinds of things. Mainly through Marc Sarrazin, who ran DeBragga & Spitler, a meatpacking district institution. He was very politically connected, as were all the butchers," he says. "Most of the gourmet societies," he adds with a laugh, "were started by the meat purveyors."

Half of New York–based French chefs had passed through Sarrazin's small office behind the cutting rooms at DeBragga's lower Manhattan headquarters. He and Yaseen agreed that this was a job for André Daguin, who already was a seasoned traveler and something of an ambassador for his region. Daguin does not dispute his role, offering a bit of restaurateur math as the explanation. "We had six Michelin stars in the Gers," he says proudly, "but when the season ended, it was the same for all of us: dead." The frankness of the region's cooking had an immediate appeal. This was no frou-frou lightness but a food grounded in farmhouses where cèpe mushrooms, goose, and foie gras were put up in autumnal rituals. Daguin had lent some nouvelle cuisine smarts to the style—meals at the Hôtel de France were punctuated with a prune and Armagnac ice cream—but he hadn't strayed far. Duck and goose confit had pride of place on the menu; the roast foie gras was served with pureed garlic that had also been simmered in duck fat.

To understand Jean-Louis's originality, it is necessary to pause at this juncture. It was impossible to be a chef from southwest France and not use those local ingredients. But was it what one would call *ingredient-driven* cooking? That is the modern ideal, a sort of truism that real quality can only be achieved by expressing what is already inherent in raw form. But this modern ideal was not Gascon cuisine, which if anything was heritage-driven and often at its best when modestly expressed. Daguin wasn't above sautéing and serving his farm chicken with whole cloves of garlic. Coscuella made a simple omelet stuffed with crackling duck skin, which he served with slightly cool local Madiran. Like the other chefs who would eventually make the Burgess trip, Jean-Louis needed local ingredients, not because they drove the menu but because without them he couldn't cook. Where he differed was that the heritage he drew from, certainly by the standards of the Gers in the late '60s when he was apprenticing to cook, was broad enough to include not only one heritage but two.

The son of a Spanish mother and an Italian stonemason father, he had grown up on the Spanish-inflected cooking of his mother's stews of dry sausage, beans, red bell peppers, and onions. It was not a particularly happy home and, already a year before his father died, Jean-Louis at twelve had found shelter in a local kitchen called Le Regent under the tutelage of its owner, René Sandrini. It was a close apprenticeship—Sandrini had sent him to École Hôtelière in Toulouse and sponsored him through various posts in hotel kitchens in Paris and Monte Carlo.

It was always Sandrini's plan that Jean-Louis come back. A devoted opera lover, Sandrini dreamed of leaving the tight confines of Le Regent—where a good prix fixe lunch was served to local workmen—and taking over the stone chapel at the monastery of the Cordeliers in the center of town. It wasn't long after Jean-Louis's return that they set about renovating it. Excavating the nave, they'd laugh over the old story that there was a passage linking the monks to the nearby convent. When the restaurant, La Table des Cordeliers, opened in 1968 with a tapestry and a soaring arched dining room, it gained almost instant acclaim. The Michelin stars came quickly. The 1974 edition of the *Gault-Millau* guide praised the *brouillade de cèpes* and the *lièvre à la royale* while bestowing a very respectable 16 out of 20 score, putting it in the good company of L'Ami Louis in Paris and immediately below legendary establishments such as Pic in Valence, L'Oustau de Baumanière, and La Côte d'Or in Saulieu, all of which garnered a rating of 17. Still, there is a note of hesitation in the praise. Things weren't quite settled. "Monsieur Sandrini is sick," the entry reads, "and we want confirmation that young Palladin, an exceptional cook, can still progress."

BILL BURGESS GOT A TASTE of Gascon hospitality one month prior to the Atlanta dinner when he went to visit the chefs on their home turf at the end of October 1978. Together with another couple, he and his wife, Joan, a TWA flight attendant, had done a tour of the restaurants. The harvest was in, and at each they'd found the chefs busy putting up the confits for the winter. The snapshots they took—Burgess resplendent in leather waistcoat and bell-bottoms—inevitably have a copper kettle used for poaching the duck legs in the background. At both Coscuella's Ripa-Alta and Daguin's Hôtel de France the work of cooking meat in these cauldrons of duck fat was done in the courtyard. At each stop there were cèpes and duck and foie gras to eat and the just-distilled white Armagnac to taste before it even went into the barrel. It all caught up with Burgess on the final leg of his journey back to Connecticut. "When we got off the plane in New York," he recalls, "I lay on the backseat of the car all the way back home to Westport."

Jean-Louis had not participated in the festivities. Condom stood far enough away from Auch that it didn't have to be included in a quick itinerary. And that suited him fine. There was trouble at La Table des Cordeliers. Sandrini, who'd always had kidney troubles, had died in 1976, and the understanding with his widow had always been that Jean-Louis and his wife, Régine, would become equal partners. That wasn't to happen. Without Sandrini at the center, the many years in close proximity they'd all spent eventually resulted in simmering tension and recriminations. It was with all this on his mind that, in early November 1978, Jean-Louis boarded the plane with his fellow musketeers, traveling to America to promote their region and their restaurants at a time of year when little else was going on.

The trip itself was enough to put all Jean-Louis's troubles at La Table des Cordeliers out of his mind. The chefs cut a swath as they traveled on the Chaîne des Rôtisseurs circuit. There were lost passports and misplaced bags. In Amarillo, Texas, at the home of the local chapter's *bailly,* a well-off dentist who was hosting a dinner, Coscuella heated a pan to such a high temperature that it set fire to the wall behind the stove. The fire department was summoned. Siding was ripped out. Afterward, bottles were uncorked with the guests as the chefs, merciless friends, ribbed Coscuella (who had a slight speech impediment) about his attempts to summon help with a timorous "*F-f-f-f-f-feu.*"

When they arrived in Atlanta, they were well into the tour. The restaurant where they would be doing the promotional dinner, aptly named the French Restaurant, was staffed entirely by the first-class dining room staff of the *France,* a recently decommissioned luxury liner. If there was a clash of cultures between these practitioners of a cushioned haute style and the band of more rustically inclined Gascons, it didn't show. A professional comradeship kicked in, and everyone's energy was focused on the event. After sending out embossed invitations, Burgess had gotten all the major bottlers to attend. Quite a coup. With the transition in the early 1920s from soda fountain to bottle sales, these men had become extraordinarily wealthy. Though bottling essentially demanded watering down concentrates and transporting it through a protected region, in many communities the bottlers represented a self-made country aristocracy. Walter Bellingrath, the Coca-Cola bottler for southern Alabama, left a mansion and gardens on Mobile Bay that are open to the public. In the butler's pantry one can see the silver tray Coca-Cola presented him with in 1940 when he joined the "Five Hundred Thousand Gallon Club." Propped on a nearby shelf is the tall silver vase he received when he became a posthumous member of the "Million Gallon Club" in 1970.

The dinner served in Atlanta would be typical of the Gascon style. Burgess had paid for several pounds of truffles that were displayed at the buffet. It wasn't a stuffy dining room; the floor was rough sisal, and white canvas umbrellas lent a beachside air to the terrace overlooking the ice rink anchoring the Omni's atrium. Stations for the food would be set up all around—Jean-Louis was serving foie gras pasta; Arnaud Daguin was helping his father with salads draped with thin-sliced confit gizzards. But in the kitchen during the preparations before the meal, all the things that the trip was helping Jean-Louis avoid came rushing to the front. André's daughter, Ariane, then studying at Columbia University, had come down to help out. She remembers a flurry of telegrams and long-distance phone calls between Condom and Atlanta about the future of La Table des Cordeliers.

From Burgess's point of view, the meal was a huge success. He recalls a moment sitting out in the terrace looking up through the building's fifteen stories—atmospherically flood-lit with green strobe lights—knowing the Coke and Pepsi people were looking down enviously from the ballrooms upstairs. For the chefs it was a slight disappointment. Sure, it was one more meal among many on the tour, but a lot of the food had gone untouched. Perhaps it had been too authentic for the bottlers. What they usually got was a big buffet with ice sculptures, prime rib, and an open bar. They'd been amused by the chalice of truffles as it toured the room—its expense was quantifiable—but confit gizzards, well, they'd barely touched them. The musketeers didn't dwell on it. There were more dinners to cook in different cities. A week later they'd be in New York—where in the spirit of the late '70s they each bought very fly raccoon coats—taking cases of wine on the train when they joined Bill and Joan for Thanksgiving in their Connecticut home. Now it was the last night in Atlanta. Jean-Louis managed to forget his troubles enough to lead the chefs in a chorus of salty French songs while they slipped around the Omni's ice rink.

JEAN-LOUIS AND RÉGINE ARRIVED AT Dulles International Airport on August 10, 1979. With them were the three sous-chefs from La Table des Cordeliers: Sylvain Portay, Larbi Dahrouch, and Jean-François Taquet. The builder of the Watergate complex, Nicolas Salgo, a financier and former ambassador to Hungary, arranged for expedited passage through customs. Stepping into the muggy heat—summer in D.C. had once been considered a hardship post for diplomats—they piled their luggage into a waiting limousine and headed into the American adventure.

When the situation at La Table des Cordeliers had continued to deteriorate after his return to France, Jean-Louis had started looking for a job away from the restaurant he had helped create and the town that had formed him. He'd spent several weeks in Courchevel, a ski resort in the Alps, but he felt it was too seasonal. Pau, a pretty town nestled in the foothills of the Pyrenees, was a possibility, until he rejected that, too. A friend put him in touch with Salgo, who was in Paris looking for a chef and who immediately offered him a job. And just as immediately Jean-Louis started to put up barriers. Salgo, who as a child during World War II had walked from Hungary to Switzerland, was not easily dissuaded. Régine was pregnant, said Jean-Louis; babies managed to be born in the United States, countered Salgo. He would absolutely not go without his brigade; visas would be secured. Finally, the dog, Tibo, a medium-sized French poodle that had been a wedding gift in 1970—no, they were not going without Tibo. Fine, instead of staying at the Watergate Hotel, where dogs weren't allowed, Salgo would provide them with an apartment nearby.

It's hard to resist indulging in what might have been Jean-Louis's first reaction to seeing the storied complex, settled among a sprawling series of buildings housing the hotel, offices, and apartments on Virginia Avenue, walking distance from the Kennedy Center and the State Department. The very word *Watergate* loomed large in the American consciousness. Among those offices were the headquarters of the Democratic Party, and when they were broken into in 1972, the Watergate scandal led to a transformation of the nation. As Elizabeth Drew, a longtime *New Yorker* correspondent, put it in her book on the Nixon impeachment hearings, "The word 'Watergate' may have undergone one of the swiftest and greatest enlargements of meaning in the history of the English language. A place became a metaphor. Like 'Waterloo.'" Now Jean-Louis stood on the parking by the hotel's porte cochere. Glimpsed through veils of heat beyond the pool, the Potomac Parkway followed the river's course. He was in his new home.

An awareness of the challenge was undoubtedly why he'd insisted to Salgo that his three sous-chefs come with him. They were a tight-knit trio who'd roomed together in Condom and supported one another through the punishing schedules just as they devised ways to amuse themselves in the infinity of time that a day off in a small French village represents. Dahrouch had started out first. The son of immigrants who had emigrated from a small village outside Tangiers in northern Morocco, he was fifteen and doing poorly in school when he'd been funneled toward a culinary career. The kitchen of La Table des Cordeliers gave him a new home. There were lighthearted moments when the three boys would be sent for firewood in the old Citroën van and invariably one would be left behind by his laughing friends. There was also a great seriousness. "You weren't allowed to ask how something was done," recalls Larbi. "Instead you watched. And when you were asked to do it you knew how to." Or didn't.

Splitting his time between the technical school in Auch and the kitchen in Condom (and occasionally cycling the thirty kilometers to visit his parents), Dahrouch internalized how Jean-Louis wanted things done. "We communicated through the eyes," he says. "First it was 'Larbi, do this.' Then it was 'Larbi, do this, that, and the other thing.'" All the time he was learning technique, Jean-Louis's technique: the slow construction of flavor, the multiple reductions, that you never threw out foie gras fat, that the golden liquid you might lift from a terrine was used to sweat the shallots and chestnuts that slowly roasted as the base for a haunting silken and rustic soup.

The cooking Larbi was learning in the Ècole Hôtelière was not the antithesis of this but a distant progenitor. Jean-Louis, too, had learned to diagram the mother sauces, that a Mornay was a béchamel with cheese and a sauce Choron a béarnaise with tomato. He'd learned the classic garnishes and used some of them, such as the fried parsley that brought a concentrated herbal richness to a platter of calamari or smelt, throughout his career. After a three-year apprenticeship, when Larbi presented himself for the Certificat d'Aptitude Professionnelle, the capping exam that would make him a professional, he was well prepared. André Daguin was one of the judges observing the apprentices cook their designated dishes in the cooking school's cavernous kitchen. That night Daguin, Jean-Louis, and Larbi celebrated Larbi's new status at the Hôtel de France. It was a celebration in the spirit of Gascony—Daguin cooked foie gras seven different ways.

Such days were a distant memory by 1979 in D.C. The site of Jean-Louis at the Watergate had formerly been the Capital Democratic Club, where the entrance was down a windowless passage from the underground parking. The room was small, enough for fourteen tables, but even so, producing from the pocket-sized kitchen was brutally hard. Jean-Louis, who in France had added an additional rustic layer to his cooking by grilling over vine shoots or smoking duck breasts in the large chimney at La Table des Cordeliers, was now asked to produce his magic with two flattops and eight burners.

Writing in the *Washington Post* on December 2 of that first year, in a piece titled (of course) "Breaking In at the Watergate," the food editor William Rice painted a hopeful picture of the situation. "Palladin will have the luxury of working as an artisan. He is not threatened with a need for mass production, nor a demand he work at an unaccustomed pace." In reality things weren't rosy at all. That first winter was hard. Régine had given birth to Olivier in December, and though she was perfectly happy to be away from the strife in Condom, she was alone in her contentment. Tibo, who had led a carefree existence wandering Condom's streets (they'd receive calls at the restaurant of the fights he got into with bigger adversaries), was miserable as a citifed dog on a leash. Jean-Louis faced operational problems. Welcoming Ben Bradlee, the legendary editor of the *Post* and husband of Sally Quinn, as "Mr. Quinn" certainly showed there was room for improvement. The greater problem was culinary. The hardship for chefs is that their research occurs in real time, with full and expectant dining rooms waiting outside the kitchen door. Jean-Louis just couldn't get traction with what the purveyors brought. Rice's profile paints an indelible picture of the famously farsighted chef at work: "Palladin's body extends forward from the waist, absolutely parallel to the floor. The sputtering sauce is only inches below his nose. A stir. A sniff. His concentration is total." But the dish he is putting out is fairly generic haute food, two ovals of poached salmon mousseline. "It's a challenge," Rice writes, quoting Jean-Louis. "Since I began cooking, I used the products of my region. Now I am here, and I will experiment. To some extent the restaurant will be a laboratory to see if I can surpass what I've done before."

WHAT HE HAD DONE BEFORE was singular, playing off the roughness and rusticity necessary in Gascon cooking and paradoxically heightening it through techniques that were the very symbols of refinement. That robust base of chestnuts, garlic, and onions roasted in foie gras fat became an unforgettable soup when it was wet with a meticulously executed clear consommé. How to get that kind of deep-rooted cooking from American ingredients, some of which he'd never seen before?

Régine dates the turnaround to his discovery of American lobster and corn. It is impossible to not think in hindsight of the famed corn soup that Jean-Louis started making as the gateway dish. He cut the kernels off the corn ear, and with the cob made a stock he used to wet a base of successive reductions—first of shallots and white wine, then of the cream that was added—before adding the kernels back to the liquid and pureeing. People still talk about the first time they tasted it. Phyllis Richman, former longtime restaurant critic at the *Post,* considers it a symbol of his cooking. "His food was always more than you would expect. It would appear to be a simple soup but it had a depth no one else could do. It was deceptively easy."

By the kind of stroke of luck a writer can only hope for, Richman was to essentially cover Jean-Louis's discovery of

American ingredients. Rice had left to become editor of *Food & Wine* shortly after he published the Jean-Louis profile, and now, as editor, and with no need to be anonymous, Richman could follow and report on Jean-Louis's discoveries. A compact five foot four, given to fashionable short bobs and printed scarfs, she had come up writing reviews for the *Baltimore Jewish Times* and freelanced for the *Post* before being brought on as staff. For more than a decade, the pages of the *Post*'s food section would be the place for reports of Palladin's discoveries and triumphs. "It could be very simple things," she recalls. "There was the guy he got eggs from that were just the right size for him. Or the moment he discovered Oregon morels." For one article she took him and a young Daniel Boulud, then a recently arrived private chef at the European Commission, to a Pennsylvania Dutch farmers' market. "Three things excited him," she remembers. "The delicate celery grown under the earth like white asparagus, the deep-fried chicken gizzards sold by the pound, and a layer cake thickly piped with Crisco frosting in red, white, and blue—pure, garish American folk art."

Gizzards! Eaten by Americans! One can imagine Jean-Louis's excitement. And this was without reporting on the zucchini flowers he got from Joan Nathan's garden, the arrangement made with a Fairfax pig farmer to drop hams in Annandale so he could pick them up, the shooting expeditions to Perona Farms in the woods of northern New Jersey, or the special light rolls he commissioned from D.C. baker Mark Furstenberg. The moment when chefs first throw off the yoke of expectations and instead of dealing with imported foods deal with local ones, becoming the center of a network of artisans and farmers, cannot be pinpointed. Already in 1979 Larry Forgione at the River Café had coined the term *free range* for the chickens he had delivered from Pennsylvania. And before that, in 1968, across the country in California's San Joaquin Valley, David "Mas" Masumoto's family had planted three acres of Sun Crest heirloom peaches with little hope of selling them because they didn't ship well, but at least it kept the variety viable and alive.

The hands that something passes through determine the way it continues. Jean-Louis found in American ingredients the basis for his soulful style. He would perhaps have resisted that description. He was a cook; the over-the-top compliment was the stuff civilians trafficked in. Humor was more his line; sometimes, when they'd have a photo shoot in the kitchen, the bespectacled chef would stand beside a commis as he cut something—the very picture of a mentor—whispering, "I can't believe how badly you suck at cooking," intended to make the young cook crack up.

But ultimately everything depended on the available ingredients, and once you proved yourself a worthwhile provider to Palladin, the demands became great. To a certain extent the ingredient is the muse. When Palladin first met Rod Mitchell of Camden, Maine, in the summer of 1980, he was running a wine and cheese shop and diving for scallops for his own use. He came from a family who knew the inlets of the state. His grandfather had made and branded his own cedar buoys that floated above the lobster traps, and his uncle had been a fishing boat captain in Boothbay Harbor and a striped bass guide on the Kennebec River. It was an expertise that Jean-Louis responded to; he took an immediate liking to the outdoorsman and the produce Rod could secure him. Mitchell remembers, "He sat me down in front of Alan Davidson's *North Atlantic Seafood* and we went through it together and we wrote out the list."

Nowhere is the connection between supplier and end result clearer than in these loose pages. Here, in Jean-Louis's elegant handwriting—a script one recognizes immediately from the menus—are listed the page numbers and names of what he insisted Mitchell secure for him. Red Mullet (Cape Cod) *Mulus auratus,* page 113. Brill, page 145. Spider Crab–Spiny Crab, page 205. Whitebait and Small Smelt get no page number perhaps because there are several varieties. All very hard to find. The moment Jean-Louis saw them he was excited. These were the *piballes,* or baby eels known as elvers. At the restaurant he would cook them with finely diced peeled red and yellow bell peppers; the sauce, embodying his rustic elegance, was lobster consommé finished with a knob of garlic butter that had minced chives whisked in just before serving. When he prepared this in Maine, he put the earthenware *cazuela* right on the campfire and simply used a splash of white wine.

"Rod, you need to take the flashlight and look in the water at night," Mitchell recalls Jean-Louis instructing. "So I had my books from marine biology and I started reading about the eel metamorphosis. They spawn in the Sargasso Sea and they float in the salt water of the Gulf Stream coming up the coast until they smell freshwater and they come up the rivers and streams of the coast of Maine early in the spring and on the high tide."

That might sound harvestable enough, but one has to hear the details to understand what was involved. Mitchell would set up in Camden, right near the falls created by the Megunticook River. "The elvers need the velocity of the tide to get upstream. They need help," he clarifies. "We used flashlights to find out where they were. You find the stream, and you wade across to dip the net. I had my wife make these fyke nets that are like mosquito nets, and when the tide went down we'd set the net at the base of the waterfall and the eels would come in."

The unrelenting nature of the restaurant business kept everything sharp. "There was a lot of craziness going on. 'I need this by tomorrow night.' He'd have me driving all over Maine to pick up crabmeat, the live lobster, scallops. I had a pickup truck and I'd drive it to Boston and pack it myself and send it off." Was Jean-Louis grateful? Mitchell laughs and puts on his friend's gravelly voice. "He swore at me all the time. He'd say, 'Rod, you need to get serious about your work. You get me that fish or no more business for you.' And he would hang up on me."

A great deal of the tension was due to Jean-Louis insisting on writing the menu each day a couple of hours before the restaurant opened. "He'd go into his office, have a cigarette, and write out forty-two menus by hand," recalls Jimmy Sneed, his sous-chef for six years. There is something in that need to see what the last delivery guy might be racing in with before committing dishes to writing that speaks to the individual nature of Palladin's cooking and, more broadly, to the word Rice had used to describe him: *artisan*. Yes, it still had enough currency to conjure up a personalized undertaking. Once though, it carried a more public connotation. In an 1862 draft of her poem "It Sifts from Leaden Sieves," Emily Dickinson paints the way a New England snowstorm makes all activity cease:

It Ruffles Wrists of Posts
As Ankles of a Queen—
Then stills its Artisans—like
Ghosts—
Denying they have been—

What did the word mean by 1980? It existed somewhere out there carrying the connotation of a colonial blacksmith's shingle but very little currency and less ability to name things in the modern day. Mass production had transformed its meaning, from the millwright or carpenter whose business might cease because of a savage nor'easter to something slightly effete, a level of quality available to those who could afford it. Reclaiming a more collective heritage would be one of the great achievements of the modern artisanal food movement, but first we'd have to face all the larger questions that bubbled underneath it. Is scale the antithesis of quality? That sounds kind of elitist. If one argues for a more generally available level of quality, is it still okay to feel a slight chill at hearing *artisan* bandied around like a catchall marketing term? In the subterranean kitchen of Jean-Louis at the Watergate, the clear progression still made the word apt. With the last delivery signed for and the menu written, Jean-Louis would pop a Mentos into his mouth and step into the kitchen. When he looked at the faces of his team, he saw the people who had learned how he wanted things done, the plating, the sear, and the tricks, like the raw lobster roe compound butter that, starting deep green and turning orange when heated, would finish a silky sauce.

PART TWO

For about a year when I was sixteen, I was a serious baker. Elizabeth David's *English Bread and Yeast Cookery* was out in a Penguin paperback edition, and I was working my way through it. This was in Dublin. I'd buy rye flour from one of two macrobiotic stores, a plug of fresh yeast from a bakery in the Liberties, the neighborhood of Jonathan Swift's St. Patrick's Cathedral and the St. James Gate Guinness brewery, and take them home. I'd mix the dough in a bowl, let it rise in the closet above the electric water heater, knead it on the top of the waist-high fridge, and pop the mixture into the old, preheated cast-iron oven. In my mind, I was detaching from technology by making my own bread; in fact, by using specific tools to accomplish my goals—from the fork I used to swirl the dissolving yeast to the serrated knife with which I cut the loaf into slices—I was being extremely technological.

Those were one serious sixteen-year-old's efforts, but in the decades since, the rejection of technology has grown into a stance. It's understandable since, once technology got braided into industrial efficiency, it smoothed the rough-edged individuality out of American food. Jams that had been thick with purple-stained berries became innocuous pastes, bread that had crackled on the bakery shelf was suddenly something diminished, bagged and sealed with a color-coded sell-by date. Today restaurants make allowances for a few next-step devices such as nitrous oxide cream whippers to aerate sauces or *sous-vide* machines that allow for precise poaching, but chefs want nothing to do with the term *technology*. The cultured butter, the hundred-year-old Puglian starter for the pizza dough, the cocktail bitters that the barman concocts, the twine-tied chalky *salumi* aging in a temperature-controlled nook visible from the dining room all say one thing: this is a place of craft.

Craft and technology, however, can't exist as contrary concepts. Many of the hand-fashioned skills we celebrate were useful forms of technology in their day. The metal hoops and wood staves of a barrel might be colorful reminders of another time, but a barrel was just an early transportable container for goods from flour to pork bellies. In an age when photos of what food we're excited about ping between our phones, it requires stubborn discipline to maintain an either-or division. When we hesitate to acknowledge the importance of technology, things can get silly fast. We've all been there: in those interiors where hoary pulleys are used as decorative settings, drowning in letterpress font pretending it doesn't come from a laser printer. It's fun to go preindustrial, getting together with friends to pedal-churn a batch of ice cream. But that's a pleasant Saturday afternoon, not a viable position.

In tracing the history of three key elements of the American table—bourbon, baking soda, and beer—I draw on certain themes such as adaptation to new conditions and adoption of new ways of doing things. Rye, not corn, is what a Maryland farmer had been distilling in the 1780s, but once he crossed the Allegheny Mountains, corn was what he had. The beaten biscuit had been an airy enough treat before the introduction of baking soda, but once that quick form of powdered leavening became known, it became immediately popular. In the 1840s, twenty years after it was introduced, baking soda was a staple for those wagon trains stocking up in Council Bluffs, Iowa, for the long journeys out along the Platte River to settle the Great Plains.

How did technology go from necessary to shunned? What were the steps that efficiency dispensed with that ultimately returned as values to be reclaimed? We can't just saddle efficiency with the blame for the loss of characteristics. Finding ways to perform the same task with a lesser expenditure of energy is one of mankind's great motivations. That applies as much to the weightless gear changer on a serious cyclist's back wheel as it did to the beautiful eighteenth-century pie crimpers of colonial America where, with one circling move, the home cook cut off extra dough and decorated her pie's border.

This then is an effort to look back on certain foods from the vantage point of today. It's an exciting time to be cooking. We can use the old metal box grater that's been hanging around the kitchen forever or we can go out and get a professional Microplane model that will render that last dusting of lemon zest or *grana* cheese feathery soft. As a critic, it excites me to see a chef command every register of technology. When Kris Yenbamroong of Night + Market Song on Sunset Boulevard in Los Angeles makes a dressing for *som tum* salad, he uses a clay mortar and wood pestle because those tools will bruise the garlic without crushing it, allowing him with each stroke to emulsify a sauce of chiles, palm sugar, peanuts, and fish sauce that he can toss strands of raw green papaya in just before serving. When he tweets a picture, he makes it part of the public discourse.

TO DISTILL

The gift store of the Maker's Mark distillery in Loretto, Kentucky, is a busy place. Crowds dawdle in the aisles, choosing between calendars, books, and rubber mats stamped with the bourbon's trademark logo. The largest group is made up of people waiting patiently in line to dip bottles in the signature red wax coating that runs down the neck of the brand's bottle, enclosing it like an embossed seal. These folks want to perform that process themselves. One by one, they're led to one of the handful of stations where the bright wax boils in a vat as a smiling lady helps them with this last handmade step. Keepsake or gift, the bottle is now individualized. The mood is one of tolerant expectation. Dressed in bright acrylics and trucker hats, these are farm people, who neither rush others nor are in the habit of being rushed themselves.

Outside the distillery is a cluster of buildings painted the same matte black as traditional Kentucky barns; open fields stretch into the distance. In some places, the flatness undulates, marking where a runnel of rushing water has cut into the soil. Here at

Maker's Mark, a creek runs right through the distillery complex. Channeled by river stones, its brackish water flows as forcefully as it must have in 1805 when a certain Charles Burks first corralled it to drive the wheel of his water-powered gristmill. This combination of the present and the vivid past is a mix I've become quite used to as I spend a few days taking the Kentucky Bourbon Trail. It's a heavily promoted industry tour—if you get a "passport" stamped at all nine distilleries, you receive a free T-shirt—but the itinerary does suggest off-the-beaten-path locations, points out bathrooms, and provides a dependable informed welcome at each stop. As someone who's tried to tour Bordeaux wine country, I don't take any part of that for granted.

I won't be getting that T-shirt since I haven't been to all the distilleries. The faux-antique-stillhouse look and polished atmosphere at Jim Beam kept me from signing up for that tour. Other distilleries I've gone to aren't part of this circuit. Willett, outside Bardstown, was closed by the time I got there late in the day, but driving up the track that leads from the looping country road to the stone buildings had its own kind of magic. Standing on the bare hilltop, looking over the bleak landscape in the dim evening light, I'd remembered whiskey's warming power and force, its mix of sweetness and astringency, the smoothness and raw edge of its kick.

Here in bourbon country, it is easy to see how basic technology was used to produce something unique to this region. A mill wheel in motion and copper's conductivity are the basic principles the process of distilling once depended on and are the distant antecedents of today's strict definition of the alcohol. For the farmers who settled the Kentucky region in the late 1700s, these streams transported whatever goods they had to sell and linked them to settlements when there were still few roads.

The creeks also made whiskey possible since they powered the gristmills that ground the corn that eventually became bourbon in Kentucky and whiskey in Tennessee. The corn was brought to these turning wheels, milled, cooked up into a light alcohol, and, after a trip through a dinged, riveted copper pot still, emerged from the condensing coil (the creek's water cooled that, too) as raw frontier whiskey. There were no nuances to the flavor of that bracing drink, no hint of vanilla or saddle leather color, though even the smoothest of today's bourbons is linked to those early days by the stipulation that the mix of grains it is made from must contain no less than 51 percent corn, be distilled at 160 proof or less, and have no additives other than the water that brings that proof down. It must also be aged in new charred white oak barrels. Aging is what rounds, softens, and adds the buttery caramel that makes smooth even the highest-proof version. To get that mellifluous finish, the barrels are stacked in rickhouses where they exhale in the summer steam and intensify their golden glow during winter's frost, concentrating their liquid contents all the time. The goal is to let cold and heat, together with the charred curved wood staves, combine to temper the edge on the raw whiskey and give it an amber shade.

The rickhouses occupy the landscape in a curious way; they are essentially depots—and accordingly some go with the utilitarian concrete look—but they are also a reflection of the individual distilleries. Located in a dip amid the white fences and perfect broad fields of bluegrass country, Woodford Reserve in Versailles is probably the most elegant distillery. It, too, started beside a stream, in 1812, when Elijah Pepper built his gristmill on the Grassy Springs branch of Glenn's Creek that today is a compound of old brick and stone buildings. The copper accents and earth tones of its welcome center (and the ten-spot for the tour, which is high for this area) are all clear reminders of this bourbon's top-shelf credentials. The rickhouse here is a gorgeous structure, fed by parallel rails that allow workers to push the barrels over from the main buildings. Standing inside the quasi darkness with the hooped staves rising toward the high ceiling in shelves as the barrels patiently exude traces of their contents is fairly magical.

But the one at Four Roses in Lawrenceburg is my favorite tour. Everything is functional here. An incongruous mustard-yellow Spanish-mission-style structure, the distillery reminds me of a resort hotel where you might get free breakfast and a decent rate. What Four Roses lacks in staged authenticity like some of the other distilleries, it makes up for in a vivid sense that you actually are walking through where bourbon is made. There are some loading bays and some workers on break. The rickhouses are nondescript buildings on the other side of the barely moving Salt River. Most of the barrels are trucked to the bottling center closer to Louisville. The tour is free.

After some welcoming remarks, my tour group climbs some stairs to the mash room, where the process begins. The scent of fermentation that has been a backdrop of sorts at every other distillery now hits me with its elemental force. It isn't so much that it rises from the four huge wooden vats filled with what looks like steamy gruel, but that it seems to envelop all of us who stand in it, saturating the atmosphere with its warm breath. Long before anything laps the sides of an oversized ice cube or puts some backbone in an old-fashioned, it starts out here. The smell is mycelial, hovering in the fungal range and emanating from the bubbling yellow cornmeal in the gigantic mash tubs like vapor from a primordial pond. "Go ahead, stick your finger in it," says our tour guide, a middle-aged woman in a company polo who seems genuinely concerned that we get the maximum information from our visit.

That's not a sentence you'll hear on most food-related tours. In a bakery—as a special allowance—a visitor might be allowed to dimple rising dough with a finger. In a cheesemaking operation, touching anything is out of the question. Those kinds of places are restraining themselves when they request you wear hairnets and paper booties. They'd like to see you in a hazmat suit. Here, it's different; the microbe doesn't stand a chance. What's drained from these tubs will have a low kick of about 5 percent alcohol, not potent enough to preserve the liquid, but good enough to get it into the columns of the copper whiskey-manufacturing still the guide shows us, where it will be heated and condensed repeatedly many times. That really is the essence of distilling, exploiting the different boiling points of alcohol (173.3 degrees Fahrenheit) and water (212 degrees Fahrenheit) to lift the ethyl alcohol like steam from water that hasn't yet reached a boil. The process is not without poetry. The Italian writer Primo Levi, a chemist by training, wrote that distilling is beautiful "because it involves a metamorphosis from liquid to vapor (invisible), and from this once again to liquid."

At Four Roses, the prosaic rules. "Is that called bourbon?" someone asks the guide of the clear liquid that emerges from the stills. No, because it hasn't been aged.

"We call it 'White Dog,'" she says, evoking with the mountainy burr that lingers in her accent any number of Kentucky hollers where only a few decades ago 200-proof moonshine might have dripped out of a quickly disassembled condensing worm. Those days of illegal distilling are mainly gone; the depth the secretive rituals gave a community are contrasted with the sedate ones of today by Lora Smith when she writes in the mixology zine *Punch,* "I'm doubtful that ordering a $7 Mudslide at the Applebee's out by I-75 lends itself to the same level of adventure." With bourbon moving toward a premium drink, "small batch" has become a useful—if not legally defined—way to describe a blend of only a few chosen barrels. When we repair to the tasting area, we group around our guide as she sets up some bottles and small plastic cups and gives us generous pours. The small batch is definitely smoother, though I think I prefer the no-nonsense kick of the regular Four Roses going down neat.

MY TOUR OF KENTUCKY HAD started a few days earlier across the state at the Cumberland Gap, a saddle-shaped opening in the Appalachian Mountains through which the earliest settlers passed into this new territory. There are two vantage points to this natural gateway for the visitor: the first is from the tiny town that bears the name of the mountainous cleft where I'd warmed up with a latte. It was early April, but thin sheets of ice still fringed the stream that meandered by the tiny coffee shop and bookstore. The second is the Pinnacle Overlook at the end of a curving road—here the thin ice became sharp, long icicles—that climbs to almost twenty-five hundred feet and looks down on the town and the entire surrounding area. I could see three states spread before me. A wide swath of Virginia farmland stretches to the east; beside it and sweeping westward lies Tennessee; to the north is Kentucky.

From here, you get to see exactly what the gap meant and why in the late eighteenth century it became a storied place. It had been used for centuries by Shawnee and Iroquois peoples as the way to travel between hunting grounds. For the immigrants coming down from Pennsylvania and inland from Maryland and Delaware and for whom the Alleghenies represented a steep and impassable front of scraggly stone and forest, it was the first point of access to new land. For these exhausted migrants coming from the crowded cities of the Eastern Seaboard, the Cumberland Gap showed them their first glimpse of new untrodden land. Up they went toward that clearing, driving heavy oxcarts that carried everything they owned. Many of the first settlers were the Scotch-Irish who made up one of the earliest immigrant waves to arrive in the American colonies. In the 1780s, their cooking utensils were high-sided red clay pots with iron covers that protected their food from ashes, and when they paused for the night and gathered around those fires, they brought out fiddles and played old country tunes like "Barbara Allen."

These immigrants also carried stills, some as small as twenty-gallon capacity, though fifty was more common, that could be dismantled and carried on packhorse or oxcart or even strapped to the back. These were people, as Henry G. Crowgey wrote in *Kentucky Bourbon: The Early Years of Whiskeymaking,* his exquisite history of American whiskey's early days, "who regarded liquor as a necessity of life. The distillation of liquor or brandy occupied the same place in their lives as did the making of soap, the grinding of grain in a rude hand mill, or the tanning of animal pelts; distilling equipment was as necessary as the grain cradle, the hand loom, or the candle mold."

Originally those stills would have been brought over from

England, but by the 1750s Frederick County, Maryland, was becoming known for its copper mines, and a rich vein was also discovered around Pluckemin, New Jersey. When Paul Revere opened the first copper rolling mill in Canton, Massachusetts, in 1801, large thin sheets that could be fashioned into many household objects became available to the nation's growing number of coppersmiths. These early American craftsmen were not interested in making decorative objects; they dealt in the everyday: riveting handles onto pans, piercing the holes of coal warmers kept under tables, hammering out the smooth bottoms of fruit kettles used for making apple butter, and sealing the seams of stills that could be disassembled and carried on a packhorse like those crossing the Cumberland Gap.

The millstones, too, were originally brought from England. They had been transported from Britain as ballast on ships, then unloaded by Virginia merchants and loaded again onto carts that creaked up the slopes of the Alleghenies. Each heavy stone represented a technology breakthrough: attached through gears and cogs to the vanes of a turning mill wheel, it was a way to grind grain using the force of water rather than of man or animal. Even with the burrs chiseled on their surfaces smoothed by age, the heft of these early tools is still palpable today, when spotted as lichen-covered monuments in a Bardstown park.

For early Americans this is still a preindustrial realm where basic, often ancient technologies were being introduced to new locations, unmodified in any way from their earliest forms. Certainly the waterwheel represents both a form of emancipation and an enlargement of possibilities. The Moravian community of Bethlehem, Pennsylvania, constructed a single mill on the banks of the Lehigh River in 1745, and it soon was supplying the force to grind wheat, saw boards, press oilseed for carpentry, strip hemp for rope, and grind the oats residents ate (and the snuff they sniffed) all while pumping water to hilltop cisterns for the growing town's waterworks.

With that capacity it is not surprising that the spread of the mill is the very marker of claimed land. Long before there was a Minneapolis—and certainly long before the city had harnessed the waters of the Mississippi to become the flour milling capital of the nation—there was a mill on its banks—and Fort Snelling guarded it. In John Filson's original map of the Kentucky Territory—published in 1784 when the future state was still part of Virginia—seven mills are shown, each named for the family that operated it. Those were still harnessed creeks, shaped by stone and wood into millraces that drove the water toward turning paddles. In the case of the Ohio River, by 1807 its banks were dotted with turning wheels, though for that great vector to the interior that rose from the confluence of the Monongahela and the Allegheny Rivers in southwestern Pennsylvania, it was often a case of gauging its quiet force so it did not take the mechanism with it.

The history of American spirits is tied to these same early technologies. The Ralston Cider Mill in northern New Jersey with its sixteen-foot waterwheel shows how local apples were ground into a pomace to be fermented into cider and distilled into applejack. The drink of farmer distillers at first, the hardy raw drink was eventually consumed at such a volume in nearby New York City that by 1810, Essex County, New Jersey, was producing close to a million gallons of applejack a year. On the banks of the Potomac River, as part of the Mount Vernon tour, visitors can walk around a replica of the gristmill George Washington had built on his property, where rye was often distilled. Both apples and rye suggested a measure of permanence no one heading into Kentucky's and Tennessee's uncharted and forested land could count on; rye implied the land was clear enough for planting, apple orchards took time to grow, time that represented the opposite of western expansion's constant pull. "I was born on the frontier and I've never lived where the apple tree bore fruit," a woodsman told an amateur historian in 1833. "We always planted them, but before they grew up, we moved."

THERE IS SOMETHING UNALLOYED ABOUT a great spirit. It's the pristine beam of light capable of refracting a rainbow's worth of colors without yielding its essence. The paradox is that this power has nothing to do with alcoholic proof, but rather the integrity of the liquid's core. That understanding lies at the heart of modern mixology. A lousy drink sloshes things together, making you feel as if you're wandering a casino floor, sucking from a bucket of bottom-shelf piña colada. A great drink is the opposite, creating an interplay between sweet and astringent, generous and austere, aged and fresh that still manages to ripple back to the ingredient from which the hooch is made. For Scotch, it is barley kilned amid heather and the splash of gray seas; for bourbon, it is corn that lends a rounded strength and fillip of heritage.

We've grown used to discussing corn in terms of monoculture or the reviled high-fructose corn syrup. But originally it was the crop of security, linking eras of agriculture and people who may have nothing else in common but their dependence on the tall crop. For the western Apache of central Arizona, who year after year planted in the same terraced fields as their ancestors, it represents the deepest of connections to tribal lands. To the Depression-era Iowa farmer, the rituals of planting were as recurrent as the seasons. In March, he would drag a pole over still-frozen ground to break up the old stalks. Harvesttime left him pensive. "I worked in the 'back forty,'" wrote Elmer G. Powers from Boone County, Iowa, whose diary is almost a calendar of life lived on the schedule of corn. "It is from this field that I can see the cemetery where my grandparents and other pioneers are buried." Planting was a moment of nerves. "Now, as fast as the planting is finished, the big worry will be, 'Will it come' 'Is the seed good.' I have been looking at our corn planting already," he wrote with some relief in May 1931. "It is sprouting."

So it was for the pioneer who set out for new locations at the first thaw and entered new territory, carrying the kernels for the next crop. Corn dictated the tempo of life, since what was planted in the early spring was harvested at summer's end, containing in each plump stalk the basic building blocks of daily life. The outer tassels that contained the cob were used for kindling, the cobs themselves to fill mattresses. Cooked in a skillet with lard, cornmeal became pone; set on a board beside the fire, it became johnnycake. At its most basic, heaped on the blade of a hoe held above the embers, it was hoecake. Ever present and necessary, the staple was an economic riddle of sorts: the more it was planted, the less value it had. The survival crop couldn't become a market crop. "Occasionally corn was exported," wrote Sam Bowers Hilliard, in a history of the food supply of the Old South, "but for the most part, it was too cheap to bear the cost of transportation over long distances and too nearly ubiquitous to enter into local trade in large quantities." Here's the paradox: the source of sustenance kept the farmer at subsistence level. To transform corn into something of even modest value, it was necessary to head to the gristmill and have it distilled.

Primary as they might be, the economic underpinnings of bourbon are rarely mentioned in any distillery. That's understandable. It's a lot more transporting to hear about old Elijah Pepper at Woodford Reserve than the fact that the distillery is owned by megascale company Brown-Forman. Certainly, when a native ingredient such as corn made contact with the basic technology of milling and distilling, something indelibly American was born. But the wheels of technology moved forward quickly. The riverbanks filled; settlements became growing towns with foundries, and ship chandlers supplied them with equipment as towpaths and taverns spread upward from the water's edge. Cincinnati as a city may have had the most notable development. In 1788, it was founded as three settlements near the Ohio River. Less than a hundred years later, in 1882, it merited a stop on Oscar Wilde's lecture tour. The Ohio River soon became a far more important gateway into the interior than the Cumberland Gap. The narrow-gauge railroads that once brought passengers to the riverboat docks were soon linked in systems growing into networks that, entering deeper into the interior, left the time of river travel behind.

Distilling entered this era of swift general progress with the development of the column still, which first appeared in Kentucky in the mid-1830s. Also called the continuous still, the mechanism consists of two columns, one called the analyzer, the other the rectifier, which represent in output what stacking many of the old-style pot stills on top of one another might have achieved. With steam fed to the bottom of the still and the liquid to be distilled entering from the top, the alcohol concentrated and grew progressively purer as it slowly made its way down each layer of perforated plates. There were enough impurities in pot-still corn whiskey that this new form of distillation was a marked improvement. But the uses of the column still soon were twisted; rectifying didn't just strip the occasional funky ester from spirits, but ridded them of all characteristics. The outcome for distillers was the ability to produce massive amounts of high-proof neutral spirits in a few hours, which were then flavored and sold as whatever the market demanded. The hint of corn, that toasty kernel at the center of bourbon, was the first nuance to go.

Real bourbon distillers found themselves underscoring their role in the tax system as a way to authenticate their legitimacy. To print BOTTLED IN BOND on your label meant a representative of the government had verified the proof and age of each barrel kept in a locked warehouse. That was not something the factory-sized rectifiers of Peoria could do. Kentucky producers went further to emphasize the slow maturing that went into their whiskey with the names they put on those same labels. Anyone leafing through an old collection can see that whether the bourbon was named for Taylor, Dixie, Crow, Oscar, Pepper, Forester, Judge, Fitzgerald, or Overholt, the crucial word in front of it was *Old*.

One doesn't have to be overly analytical to interpret such brand names as stories that link the contents of the bottle to something resonant of the American frontier. Perhaps a crusty old character, but gloriously real. It isn't that different from any food artisan's website today, which usually begins with a story you can click on even before the product or Google Maps location. *Childhood friends, visiting the old country, second career, long fascinated by—* these websites have their own cadence as they seek to make a personal connection that encourages us to learn more. That said, we kind of do want to know how Darek Bell and Andrew Webber of Corsair Distillery in Bowling Green, Kentucky, "hit a snag while working on a prototype bio-diesel plant," took a detour into hooch, and ended up making Buck Yeah, a seasonally available buckwheat whiskey.

By the end of the nineteenth century, the real distillers of Kentucky and Tennessee were playing defense in a world where one Ohio rectifier appearing before a congressional commission could blithely announce, "There is no such thing as a brand of spirits." To him, it was just a case of putting different labels on what was essentially the same stuff. The ham producers of Virginia were faced with their own form of corruption. Instead of the patient curing of razorback hams that had required salting and then fitting a hook through the hock bone to hang the haunches in the smokehouse, producers of goods such as the quickly popular Brown's Condensed Smoke instructed customers to use a sponge to apply their product. "One or two coats, and the smoking is done."

Though it's tempting to portray such people as villains, it doesn't quite do justice to the world they inhabited. A larger country was starting to take shape, built around the network of railways. No longer bound by following the general north-south course of rivers, trains transformed the axis of trade: raw materials could now be shipped west to east as railroad tracks became the dominant organizing principle—in fact, in 1883, the powerful railroad trusts carved the country into the four time zones we know today in order to standardize train schedules. Perhaps there was something self-fulfilling about all these changes. The abstract concept of efficiency that had shadowed technology since the invention of a turning waterwheel had shifted. Now there was always another way to profit, another method that used up less energy, one that took less time, that got the good to market faster, that shaved some cents off the cost. Such is the nature of technology itself.

As the map of the United States was gradually covered by capillary-like tracks, foods of all types had to find their place. Speed, in certain areas, allowed for a seasonal understanding for the first time. The Camden & Amboy line that ran along the Delaware River was known as the "Pea Line" for the promptness with which it could bring the most delicate produce to discerning Philadelphians. Cities became terminals to process and push food out into the national market, none more so than Chicago. In Upton Sinclair's 1906 muckraking novel, *The Jungle,* the Lithuanian immigrant Jurgis Rudkus has barely set foot on Chicago's Dearborn Street when he becomes aware of "a sound made up of ten thousand little sounds." What he's hearing are the herds of cattle and swine being driven toward the abattoirs in which Rudkus, like most immigrants, will work, feeding the carcasses of "the hogs which had died of cholera on the trains" into the packing plant's grinders.

A new form of food was appearing—one that was separated, seemingly forever, from what had been, that stretched to a breaking point its bonds to any locale or tradition. Bourbon, good bourbon, glowing in the glass as evocatively as an oil lamp, is still a link to the young rural country that existed before everything changed. It's an illusion powerful enough to ensure that every year thousands of Kentucky Bourbon Trail passports are redeemed, testaments of sorts to having seen the creek, stuck a finger in the mash, gazed at the rickhouse, taken in the broad Kentucky landscape. Proof, in that red-wax-sealed bottle, that one was there. At the dawn of the twentieth century, the essential connection between place and product was fraying. The whiskey that had gone downriver stacked in barrels on the flatboat's deck was now heading to market on flatbed trucks. On the killing floors of Chicago, the changes were all too dramatic; in the country's markets and crowded streets they were becoming mundane. Items were produced in distant places, and trains transported them through shunting yards and depots, their slipstreams carrying a new sense of efficiency—and unease—into the kitchens of America's homes.

TO BAKE

The changes that came with industrialization offer the first glimpse of a world we very much live in today. Producers weren't producers in the way we mean it until food was a commodity that could be easily transported; consumers were certainly not a disembodied entity—the consumer—until they began to purchase the goods the system produced. The home kitchen was where the comforts of the traditional met the inescapable fascination with the modern. This was not a new conflict; cooking ideals are constantly shedding arcane rituals. That's a complicated way of saying my parents and their friends eagerly gathered, forks in hand, around their Scandinavian-designed fondue kit; I may invite people over for that if we really get into the throwback mood and wear turtlenecks, too. Beginning around 1900, what was wholly new was that change was based on something unknown, the standardized product. Up until then, gelatin had started with a veal knuckle; marshmallows, an increasingly popular garnish, began with gum arabic, egg whites, and powdered sugar. Those cut-glass containers of garnishes that contributed a tart element to a well-set table such as spiced apples, crab apple jelly, pickled pears, or stewed gooseberries were the work of the woman of the house or, more likely, her cook. They certainly weren't goods from a brand.

The sense of unease that accompanied the shift is reflected in the journey that baking soda made from mistrusted novelty to cupboard staple. When we look at the familiar orange Arm & Hammer box with its claims of being "The Standard of Purity," we're looking at a memento from those early days of food technology when qualms had to be overcome long before a product's usefulness could be embraced. Still today it is fairly remarkable all the things baking soda can do. Dissolved in a glass of water it calms an uneasy stomach; poured into boiling water it can resuscitate even the most tired green bean (a restaurant trick); mixed into a paste it is a salve for bee stings—or a rudimentary toothpaste. And of course, the reason it's in the fridge in the first place is that it kills odors. But it has had the most impact in the mixing bowl, where a few sifted teaspoons of the chalky white powder transforms some scoops of flour and a splash of buttermilk into a dough that, without any proofing or waiting, becomes an American standard: the biscuit.

Until baking soda came along, women who wanted to stretch a meal with something starchy and warm would have to maintain a culture of yeast in the form of a starter, a long process that involved cooking up grated potatoes with a piece of already active leavening; some recipes called for hops, too. Keeping this going could be a hassle, to say the least, especially if you lived in an area subject to climate extremes, as yeast goes dormant in the cold and can exhaust itself in the heat. So when baking soda came along, its attraction was as immediate as the controversial issues of convenience versus self-sufficiency it brought up. Here was an ingredient that could not be created at home, had no regional affiliation, and, certainly for the early users, no associated memories.

The fundamental understanding of this form of baking had been brought over in the late 1700s by Dutch bakers who had learned to leach wood ashes through flowing water and reduce it to a powder form called *pearlash.* Using a by-product of wood is not as strange as it might sound—many components of early manufacture, from the pitch used to caulk boats to the potash that went into glass and soap making, were derived from felled trees. In the Mediterranean basin, lye-cured olives depended on the same ingredient. By the 1850s, technical developments and the discovery of deposits of bicarbonate of soda that could be mined led to new forms of obtaining the white powder and a new name: *saleratus* or, more commonly, baking soda. The speed with which it became popular says something about the ease it delivered. Making biscuits relies on a simple chemical reaction: an alkali reacts with an acid to create carbon dioxide, which lifts dough in the oven. It's the same process middle schoolers rely on to make the science-class volcanoes do their thing, a few spoonfuls from the Arm & Hammer box interacting with a splash of vinegar. Except with biscuits, buttermilk is the acid and the dough keeps the pyrotechnics hidden.

But baking soda wasn't the only white powder that was altering the home kitchen. When bicarbonate of soda was mixed with a powdered form of acid called cream of tartar it made buttermilk unnecessary. At first the compound of the two powders was done in the home kitchen, but eventually manufacturers started to produce what would be called baking powder. Sticking a teaspoon into a tin of this to lighten, say, a batch of chocolate chip cookies is so full of warm associations for us that it is almost incomprehensible to think that this development could have once been fraught with anxiety. For Mary J. Lincoln, the author of the best-selling *Mrs. Lincoln's Boston Cook Book,* it was the line in the sand.

One can almost hear the derisive snort as she concedes that the use of baking powder was "a convenient form adopted by many people who think it hard work to make yeast bread." Yet Mrs. Lincoln's style is such that she can both reprimand and educate. In the next breath, she also instructs the reader on the safest ways of procuring it. "When your druggist or cook is not to be relied upon, use a baking powder which has been tested and proved pure. Pure baking powders are soda and cream of tartar mixed by weight in the proper proportion, and combined with rice flour, cornstarch, or some harmless ingredient to insure their keeping . . . Soda is also neutralized by sour milk or lactic acid. This is economical, particularly for those who have pure milk and more than they can use while it is sweet. But milk is often adulterated, and, in winter, grows bitter before it sours; and the degree of acidity varies so much that the result is often failure."

While Mrs. Lincoln could sometimes seem at a remove from the domestic pressures so many women were feeling, in using the word *adulterated,* she is completely of the times. Grocery shopping by the end of the nineteenth century required a vigilance we can only imagine. That olive oil was blended with cottonseed oil and honey was made from glucose, that cherries were kept red with aniline and peas green with copper, and a great deal of pepper was ground olive pits, these were simply additional threats on top of the worry that the druggist mixing your baking powder might not be honest. Far more noxious were the embalming products, all curiously attached to the suffix *-ine*—such as frezine and preservaline—though it was formaldehyde that made swill milk drinkable or half-rotted raw fish and meats not give off a putrid stench.

Securing laws that would guarantee the safety of the food supply was clearly necessary, though it came at a cost and was a hard-won fight for the federal government. In an incisive study on the evolution of food fraud, British author Bee Wilson draws a difference between the battles for food safety as they occurred in England and the United States: "Everything was on a grander scale, and the battles between the swindlers and the purifiers were bigger too. Whereas the English fight had been one of science against science—and often seemed like gentlemen against gentlemen—in America, it was a fight of commerce against commerce." There isn't a fiendish undertone to this; changing the situation so consumers were sure of pure—or at least not rotten—food was a requirement. The journey the country would now embark on replaced what was scooped right from the barrel with smaller packages that were sealed for security and produced by large companies, introduced purity as a concept, and skillfully spliced into it the notion of brand awareness. It was a momentous shift that required a change in the role of the government, and all the energies of one particular man.

THE 1906 PURE FOOD AND Drug Act is also known as the Wiley Act, after Harvey Washington Wiley, the Indiana native and onetime professor at Purdue University who became the chief chemist of the U.S. Department of Agriculture in 1882. President Lincoln had named the first person to that role two decades earlier, which had helped with a certain amount of pure food laws. Curiously—or tellingly—in a nation rife with the basest forms of contamination of foods, tea was chosen as the first staple to be investigated, and the Tea Importation Act was passed in 1897. Wiley had a broader vision though, and even as he gained the backing of several senators, many others fought his reforms. "Pure food measures," he wrote in his memoir, "were smugly looked upon as the work of cranks and reformers without much business sense."

Understanding that the only thing capable of changing the mind of a politician was pressure from constituents, he became a scientist of the most populist strain, who burned with the passion of the reformer. He relished the word *adulteration,* including it in the title of his easy-to-understand books *Foods and Their Adulteration* and *Beverages and Their Adulteration*. Wiley's preferred domain was the grocery shelf and the drugstore counter, and his great insight may have been to primarily address the concerns of women. He branded the manufacturers of suspect nerve tonics (fraudulent medicine for typical ailments) "a fraternity of fraud and hokum," and he invariably explained science in layman's terms. He likened the widespread use of frezine to "tying down a steam valve on a steam engine" so the olfactory alarms the smell of rotting meat would have ordinarily set off were deactivated, no longer warning the shopper about imminent danger.

Meat, however, was not Wiley's main concern. He was savvy enough to understand that the great success that met Upton Sinclair's *The Jungle* when it was published might be the ideal vehicle to speed up political foot-dragging. Having the head of the Armour meatpacking concern fulminate against Sinclair was good; hearing the socialist author was lunching with Teddy Roosevelt at the White House even better. Roosevelt was still ticked off at the spoiled canned meats troops had received eight years earlier during the Spanish-American War, and the president started to relentlessly pressure Congress so the Wiley Act and the Meat Inspection Act were passed together on the last day of June 1906.

In his writing, Wiley was smart enough to bypass large city newspapers that might shorten his opinions and instead wrote at length in popular periodicals as influential in the heartland as the *Emporia Gazette* of Kansas and as popular as *Ladies' Home Journal.* His stories about the so-called Poison Squad of young men to whom he fed different doses of adulterants to monitor the results were intended to tug at the maternal instinct. He recognized the growing political power of women's organizations and rarely passed up an opportunity to speak before a gathering. "The table syrups of this country would be vastly improved if glucose were eliminated from their composition, and there is substituted for this mixed mass the pure products of the maple grove, the sorghum field, and the corn field," he says in a lyrical passage of a speech delivered before the National Congress of Mothers.

Even as Wiley campaigned assiduously for reform, his was a nuanced tone, always careful not to overshoot the mark of his audience with scientific jargon. Those vials and small bottles sold as remedies for the nation's stomach ailments, dyspepsia, and bouts of mental exhaustion didn't merit a direct attack, but their ridiculous names did leave an opening. Wiley didn't question the various infirmities; he just suggested that Dr. Bouvier's Buchu Gin of Louisville, Dr. Sherman's Peruvian Tonic and Systematizer of Des Moines, or even the succinct U-Go that Fritz T. Schmidt & Sons of Davenport was selling might not be the way to cure it.

Yet on the question of chemical leavening with baking soda, Wiley was strangely reserved. The chemist in him couldn't bring himself to level an accusation that science wouldn't substantiate.

Alum, an acid powder that was often substituted for cream of tartar in baking powder, had come to symbolize imminent danger according to companies like Rumford Chemical Works of Providence, Rhode Island, who'd been promoting its alum-free baking powder as "commended by the most eminent physicians for its wholesomeness." Mrs. Lincoln warned that if not mixed in the proper proportions, the aluminum residue left after baking was "injurious and to some extent poisonous." But Wiley, the actual chemist, was unconvinced and called for an objective and unbiased study.

Facts did not hinder New York's Royal Baking Powder Company from warning to never allow alum powders into the home kitchen. "Remember this," reads an ad from the turn of the twentieth century, "and insist on getting 'the Royal' in cans." This is a telling document not only because it marks the end of the era when the grocer might still measure out dry goods for individual customers but because it draws the manufacturer's true agenda out from its cover of healthful concern. The vigilant woman should trust a brand more than bulk, and if she wanted to do right by her family, she would reach for that brand's sealed box.

This was a significant shift. It began the transfer of authority that would take place from women armed with index cards, torn open envelopes, the back of bank deposit slips (the many places where a recipe shared between friends was written down) to corporate entities armed with seals of approval and experts with colorful names like Marian Manners and Prudence Penny, each pushing industry-devised recipes on a remarkably recalcitrant public. Writing as late as 1971 about her Mississippi childhood, Eudora Welty described how foreign it would have been to reach for a ready mix when an occasion called for special baking. "I daresay any fine recipe used in Jackson could be attributed to a local lady, or her mother—Mrs. Cabell's Pecans, Mrs. Wright's Cocoons, Mrs. Lyell's Lemon Dessert. Recipes, in the first place, had to be imparted—there was something oracular in the transaction—and however often they were made after that by others, they kept their right names. I make Mrs. Mosal's White Fruitcake every Christmas, having got it from my mother, who got it from Mrs. Mosal, and I often think to make a friend's recipe is to celebrate her once more, and in that cheeriest, most aromatic of places to celebrate in, the home kitchen."

If there's something heroic about that kind of resistance, it is tinged with the knowledge that, except in the most special occasions, it is unsustainable. We might feel a fondness for Grandmother's giblet gravy in the china gravy boat brought to the table at Thanksgiving, but on a typical day we will reach for the bottle of ketchup rather than ever think of cooking it up ourselves from scratch. As Laura Shapiro writes in *Perfection Salad*: "This was the era that made American cooking American, transforming a nation of honest appetites into an obedient market for instant mashed potatoes."

We can trace those changes most clearly in the receipt books that women often kept for private jottings and the charitable books that gathered recipes for good community causes into book form. The receipt books, the more intimate and personal archive, have cracked spines and discolored covers. They are a glimpse of the person who sat at the kitchen table: their pages are filled with penmanship exercises, snatches of Scripture, or the purl and stitch count of a knitting pattern. Sometimes even a snide aside: "Mrs. Edwards at luncheon, though uninvited." Looking through them a century later, one sees documents that once caught a housewife's curiosity. Here is a collection of ten leaflets put out by Jell-O, explaining the molded salads the product could be used for; here are tinfoil wrappers of Fleischmann's yeast, proofs of purchase that could be redeemed by mail for various parts of a silver dinner set.

The charitable books represent something more solid. These were the recipes that a woman proudly put her name to and could expect to get some feedback on from her neighbors. Usually the cause had something to do with a particular church. (My favorite dedication, published in 1911, was for the old Congregational church at Buxton Lower Corner, Maine, "which the Dorcas Society has painted, decorated, shingled, carpeted, cushioned, kept in repair, and provided with new hymn and service books.") But whether the cause was a new organ, a new roof, or the welfare of orphans, the broader story the books tell is always the same. In a remarkably short amount of time, from about 1890 to 1950, the food that Americans projected as the ideal in the public sphere undergoes a series of fundamental changes. Some are benign, perhaps even necessary. How long could we have gone with measurements like "a piece of baking soda the size of a pea"? But read enough of them and the trajectory of a loss of individuality, custom, and any sense of seasonality becomes all too clear. There's a narrowing of choices. Table relishes that began with a bushel of tomatoes gave way to the branded ketchup tossed into a cocktail sauce. Recipes that jumped with vitality, such as the butter gravy of powdered crackers and simmered dandelion greens suggested by the Women's Christian Temperance Union of Vermont to garnish boiled veal in 1895, give way to "five-can casserole" by 1955.

IN WHATEVER FORM THEY ARE collected, there is something poignant about regional recipes because we see the principles they stand for hanging on and then disappearing, seemingly forever. The packaged, wrapped, and sealed triumphed. Any vestiges of a more local tradition—say, the ricotta and mozzarella sold from buckets in the neighborhood *latterias* of Staten Island—might as well have been a concession allowed by the power that had dictated the terms of surrender. There were some flare-ups of resistance. Home gardening and pickling became little movements in the 1970s, but processed food was the nation's daily fare, a vast realm of shelves and freezers stocked with sugary cereals, an entire world of crackers and frozen dinners.

Still, I am reluctant to brand this food as mediocre or insipid. First, because the perfectly crisp Chips Ahoy! cookies in the yellow jar on the yellow Formica counter of the 1967 suburban ranch house is the stuff of warm nostalgia; second, because a change in eating habits reveals the underlying social elements at work. We couldn't go from agrarian to urban and eventually suburban, from homemade yeast to microwaves, without altering the way we eat. But most important, to prudishly accuse a realm of foods of not being fully flavored seems blind to the thrust of technology. Mechanical advantage, the principle of getting more force than you need to put in to produce it, brought us the lever-lowered olive press of antiquity and the rigid-frame cider screw press of early New England. As those principles of efficiency coursed forward into America's twentieth-century kitchens, they rendered obsolete set pieces of Americana that were, in reality, feats of endurance.

The romance of canning without electricity, for example,

would be lost to the 1925 Texas Hill Country woman who remembered the unrelenting bounty that needed preserving at the height of summer and through the early fall, when the merciless sun beat down on the corrugated metal roof of a kitchen with a roaring wood fire. "You got so hot that you couldn't stay in the house," she told historian Robert Caro. "You ran out and sat under the trees. I couldn't stand it to stay in the house. Terrible. Really terrible. But you couldn't stay out of the house long. You had to stir. You had to watch the fire. So you had to go back into the house."

With the rural electrification projects of the mid-1930s, modernity was about to become much more widespread. Unlike Europe, where compact villages could be wired with some efficiency when electricity became available to the masses, the rural farms of America were too spread out to make it worth the while of utility companies to put them on the power grid. In many cases, it was farmer cooperatives with hundreds of member owners who provided the money for the lines to reach their farms.

The design of the American kitchen took shape around the power outlets. Wiring plans created the U-shaped kitchen with the stove and fridge across from each other; another layout had the two appliances at opposite ends of an L. The sink always faced the window, with outlets for the iron, washing machine, and exhaust fan beckoning along the walls.

The long chain of modern technology that had improved daily life with its oil lamps and galvanized metal icebox that didn't rust (a precursor to the fridge) could not be stopped just because it was suddenly closing in on native flavor. In some cases, progress was the very means by which certain foods became iconic, as methods and ingredients met in new combinations. Historian

William Woys Weaver points to the fast jolt of heat provided by the cast-iron stoves that were coming into popularity in the end of the nineteenth century as the ideal means to raise biscuits that called for a few teaspoons of baking powder, which was becoming increasingly available. As important as any acid-alkaline reaction, the biscuit became an American staple simply because cast iron started to be used in stoves.

But cellophane, the bar code, the refrigerated eighteen-wheeler backing into the loading dock—it's impossible to extrapolate flavor from any of these. As efficiency became convenience and regional traditions evaporated, our understanding of food shifted. In John Cheever's 1953 short story "The Sorrows of Gin," the protagonist, a dutiful passenger of the suburban trains to Grand Central, has a moment of insight about all the foods of his business trips "that tasted of plastics." This is a new tone. Processed food was safe, but it was distrusted. It was pure yet perceived as corrupted in a way Wiley could never have imagined. It was not poison now, but it was somehow toxic.

It's not surprising that a generation of artisans chose to step away from this onslaught to make things with fewer, better ingredients and more care. The question is why—was it flavor alone or because fashioning the best version of those things we consume gets at the soul of being human? That would have sounded pretentious to the home-brew clubs with their mirthful names and their willingness to rediscover arcane technologies that started to pop up. Flavor, for their members, was magnified when you made it yourself. The sudsy six-pack from the convenience store cooler was never going to bring you as much pleasure as the mead you'd siphoned from a bucket on your kitchen counter into sterilized bottles ranged on the linoleum floor.

Though the revolt is often put in terms of a granola generation's snub, it is darker, tinged with paranoia, secure in the knowledge that the food industry had devised systems to keep us at a remove from what we ate. A 1970 examination of the expiration codes found in the modern supermarket that first appeared in the *San Francisco Bay Guardian* and was reproduced in the *Los Angeles Free Press* a few weeks later conveyed in cryptic detail the many ways consumers were kept ignorant about their food. "Pack date is usually indicated by a number between 1 and 365. By this code, 231 would mean August 19, except some manufacturers use the dates backwards so August 19 is 134" was a typical confusing technique that Helene Lippincott unearthed during her six-week study. There are two full pages of obfuscating techniques, and they are far more devious than the different-colored plastic bindings on a bag of English muffins. If the investigation sounds slightly overdone—just how badly do you need to know how long the cottage cheese has been on the shelf?—it illustrates a greater truth about the arc of efficiencies that had led to this point. A woman, an American woman like Helene Lippincott, one who had been wooed and instructed, one who could reasonably have been expected to endorse the food system that was presented to her with purchases, was rejecting it instead.

TO BREW

The guide, a dark-haired man in his midtwenties, of the Miller Brewery tour out on State Street in Milwaukee, has set quite a pace. The stragglers at the gift store where we've congregated barely have time to put on their cheesehead hats and pose for pictures before we are ushered into the auditorium. There are three families and four service members on leave taking the tour with me. Unless the soldiers are home brewers, no one is going to be popping a question about decoction or two-row barley like the ones you hear on a microbrewery tour. We're led into a plush midsize screening room and taught about the myth of "the girl in the moon," the Miller High Life logo that has graced tap handles, mirrors, bar mats, trays, and coasters since being introduced in 1907. We learn how Frederick Miller emigrated from Germany in 1854 and traveled through the United States, carrying the yeast he eventually put to use when he started the Plank Road Brewery in a single clapboard building on this very site.

Now Miller is a massive complex, perhaps not as big as the Guinness brewery in Dublin, which is a minicity, but large enough to run for hundreds of yards on both sides of a public street. Sliding from the past to the present, we climb flights of stairs to get a better vantage point of the massive brewing room with polished steel kettles that look like giant Hershey's Kisses. From behind sound-deadening screens, we gaze at the automated bottling lines that feed cans and bottles in a crazy synchronized display of sealing, labeling, and enclosing in six-pack, twelve-pack, and party-sized bundles. Taking us to yet another building along the State Street complex, our guide leads us into a storeroom the size of many football fields. This is more than floor to ceiling; there are avenues and side streets lined with beer for the forklifts to operate in. This guide is good. Because the sight might suggest the beer just sits there until it's shipped, he quickly assures us that if we were to come back tomorrow, we wouldn't see the same product. "No skunked beer here," he says as we watch a supersized forklift zoom a pallet to the loading dock.

Activity is good. Trucks are good. Massive storerooms are invigorating. The previous day I'd walked in a slowly blossoming funk through the mostly empty and abandoned remains of the former Pabst Brewery at Ninth Street and Juneau Avenue, a pleasant forty-five-minute stroll away toward downtown Milwaukee. Once comprising twenty-eight buildings spread over seven city blocks and constantly running on three shifts, the home of Pabst Blue Ribbon is a stately stone hulk where the storied name is still attached to the old malt elevator in huge letters on a scaffold that stretches across the street—though beer hasn't been made here since the plant closed in 1996.

In 1862, German-born Frederick Pabst, a steamer captain who worked on the Great Lakes, married Maria Best and thus joined one of Milwaukee's first great brewing families. The brewery's name would be changed to his in 1889—the words *Blue Ribbon* and accompanying sash on the labels followed six years later. By then, the company was large enough to open the Pabst Hotel in New York and to fund theaters in Milwaukee. A few decades earlier it had started on a far smaller scale and with a hyperspecificity that would make any modern food artisan proud. An 1845 advertisement in the *Wisconsin-Banner und Volksfreund* conjures up a tiny undertaking. "Best Brewery, Distillery & Vinegar Refinery on Prairieville Street, South side of the summit of the hill about Kilbourntown," reads the notice before proclaiming, "We wish to announce to our friends and well-wishers that we always have good bottom fermentation beer as well as good corn whiskey and vinegar for sale by the barrel."

Today Pabst Blue Ribbon, or PBR, is a sudsy brew with just enough of a tingle of hops to accompany—without challenging—a platter of Hunan-style shrimp sizzling with wok-fried peppers, a handful of whole roasted pork with crackling skin, or a pudgy bleacher frank. Snapped open in cans, pulled from taps, filling pitchers in crowded bars, it represents an ironic decision to drink what's widely regarded as low-rent beer. It may be nostalgia, but nostalgia for a time its primary drinkers never knew. The enthusiastic trucker-cap and good-jukebox PBR atmosphere is one more engaging manifestation of America's love affair with lager beer that can be traced to particular cities such as Cincinnati, St. Louis, Brooklyn, and Milwaukee. That bottom fermentation beer the Best ad mentions represents German beer, the fork in the road that led away from top fermenting ales. Hardier, less demanding to make, based on an easily cooked-up wort (the still-unfermented beer akin to a roasted-barley tea) whose fermentation was started with the sloshed-in suds of a previous batch, those ales had been the beer of Dutch New York, colonial Virginia, and the taverns clustered around Philadelphia's Congress Hall.

This later arrival—translucent, sparkling, foaming, and light in alcohol content—that would become the stuff of beer halls and barbecues depended on cold. *Lager* is short for *lagerbier,* or beer kept in a storehouse or cellar that's cool enough for that bottom-fermenting yeast to act. German brewing laws prohibited the manufacture of this beer in the warm summer months, specifically between April 23 and September 29. In the United States, things were different. Not only was a medieval injunction unenforceable, but it seemed to miss the very spirit of the thing. Beer—cold American beer—was what was quaffed in the ballpark, at the choral society's annual picnic, at the card game enjoyed by friends in the shade.

The amount of ice needed to produce it was enormous. In 1880, Captain Best required sixty thousand tons of ice per year, and Miller used hardly any less. It was an industry unto itself, with special ice teams who worked the lakes and ponds, marked, grooved, sawed the ice, towed it to shore, and hauled it to the brewery. It is professional pride that shines from the faces of the ice-cutters in the National Museum of American History where they pose bowlers askew, waistcoats buttoned, the leather-handled cords of their hauling slung rakishly over their shoulders. For beer, ice is one of many technologies that the brewing industry used to re-create the weather conditions that had existed in Germany. Floated in metal boxes on the surface of the wort as it became beer, packed in amid the ruh-casks where impurities settled out of the beer and alcohol increased, banked around the barrels awaiting delivery, ice was an indispensable part of lager production.

So when our Miller guide tells us the tour will culminate in the brewery's ice caves, I feel my pace quicken. We're led inside a stone-faced structure with a heavy wooden door built into the hill, the same hill that the Plank Road Brewery's cellar was dug into well over a hundred years before. Inside the cool dark recess, the walls are thick; blocked-off arches suggest the network of chambers—sixty miles, we are told—that represent the brewery's core. Decorating the walls are a few leather harnesses the horses might have used to pull ice from ponds and a mural of a nineteenth-century beer garden. We all feel something. The families are taking turns with their cameras, and two of the soldiers are posing by the hologram of Frederick Miller as a boulevardier that appears on a wall to say a few words of farewell.

I step back slightly to take the scene in. The modern microbrewery tour wouldn't have any of this slightly schmaltzy appeal. The brewer would talk about yeast strains and the tour might end at the bar before a chalkboard announcing what's on tap and the International Bitterness Units for each beer. Beer has become something of specific detail where technology—in a microbrewery the bottling line would be skirted not celebrated—is guardedly watched. That's an unease that's broadly shared. I can picture chefs who, in a ten-seat restaurant, make the after-dinner coffee by weighing the water on a digital scale and time the brew on an iPhone, yet would tell me that they were doing the opposite of technology because they were creating a maximum of flavor. I can work in my kitchen with its silicone gloves, mixer, Cuisinart processor, digital thermometer, microwave, and Teflon-coated pans and maintain it's really very low-tech. As if the addition of a Keurig coffeemaker is what would put it over the edge.

Standing in the dark Miller ice cave, I recall some of the ancient and resonant forms of food technology I've seen. That Ojibwe canoe in the National Museum of the American Indian in Washington, D.C., made waterproof by hides, its lightweight buoyancy allowing wild rice to be harvested and transported across the Minnesota marshlands. The rice fanner displayed in the Charleston Museum that, using braiding techniques brought over from Africa, transformed a sheaf of rush into a tool pliant enough to winnow and sturdy enough to scoop Lowcountry Gold rice. If you parse it, today's technological advances represent the opposite of this, not tradition or heritage, but seeing the hand that once had learned a skill relegated to reach for the shrink-wrapped package. This cave is a historical reenactment, but back in the brewery's heyday, with the massive slabs of ice being slotted into their spots, it was working at maximal efficiency. How and why would that lead to flavor being lost?

"All right now, who's ready for a Miller Lite?" our cheery guide asks, inviting us to the ersatz beer hall next door where we'll enjoy some samples. We're all happy with the news. A cold beer sounds good. As I step back into the daylight, I give one more thought to Captain Pabst. The small brewery specializing in pre-Prohibition beers slated to open at the original site would give a nice nod to the animated one he'd created. But it wouldn't change how things had ended. The previous day, as a grim coda to the Pabst tour, I'd slipped into the chair where he'd overseen the business. He'd sat there, in a room with a bay window, watching the horse-pulled drays stacked with cooled barrels heading downhill to replenish the kegs in the city's bars and saloons, sanguine that growth signified progress and that technology would keep him in the forefront of American brewing.

WE NEVER HEAR ABOUT REFRIGERATED bourbon cars. That's because the very idea of aging in barrels is predicated on a divergence of temperatures being good for bourbon, first heat and then cooler winter temperatures that allow the drink to concentrate and reach its essential flavor. Beer was different; the alcoholic content was not as high as to preserve it, and when beer reached elevated temperatures, it gained a dullness that chilling it at the point of sale couldn't mask. The archives of the Pabst Brewery are replete with disagreement on the amount of cold required for beer. The railway company couldn't charge for ice as cargo, so it always wanted to transport less of it. Eventually the amount of nine hundred pounds per boxcar was reached and settled upon between the breweries and the railroad.

The problem from the brewers' perspective was clear. Even with the speed of rail and the network of icehouses to replenish the cooling blocks, they could not get the beer farther than the last drop of melting ice. Growing use of a device called the Carré ammonia-compressor throughout the 1870s allowed brewers—Brooklyn's S. Liebmann's Sons is credited as being the first to use it in America—to draw on a constant supply of manufactured ice, and, a decade later, they grew to trust mechanical refrigeration long before the meatpacking industry did. Still, what brought beer from kegs to bottles, what led to bottling lines and the certainty the beer would be shelf-stable even if not chilled was pasteurization.

The word alone is enough to sound as a kind of warning in gastronomic circles. Pasteurization is the roadblock to natural microbial maturation, and because of that, it is considered the enemy of nuance. The broader argument is that pasteurization was never only about the safety of foods. In her extensive writing on the subject, Massachusetts Institute of Technology anthropologist Heather Paxson makes a forceful argument that "the introduction of pasteurization had as much to do with consistency, standardization and economies of scale—with *market* concerns—as it had to do with safety."

That reduction of flavor to a dull middle is precisely the ding on big beer. Hopped, though in no way hoppy, lacking the bright bitterness that cleans the palate as it refreshes, and shippable to the farthest convenience store's cooler, big beer in no way represents the heritage of full-flavored American beers. Plenty of craft brews are pasteurized, of course, and cost-cutting and business consolidation probably played far more important parts in the sad trajectory; still, pasteurization serves to focus the accusation—it is the process that muffled the nuance and did away with regional differences—that technology is what made our beer boring.

It is a curious charge to level at the man whose name is part of the word. Born in Dole and raised nearby in Arbois, the two largest cities of the Jura region in the French Alps, Louis Pasteur traveled to Paris in 1843 as a twenty-one-year-old to pursue a doctorate in physics and chemistry at the École Normale Supérieure. From the beginnings of his scientific journey he remained connected to the flavors of his upbringing through gifts his father, an Arbois tanner, would send him. This is Pasteur before myth, not yet the savant who fought anthrax and rabies but a son of the Jura you might engage on the stinky glories of a good ripe cheese. The gifts have all the warmth of once-shared foods sent to a distant loved one. In the lifelong correspondence Pasteur and his father maintained, there are letters from Pasteur thanking Jean-Joseph for a ham, or notes from Jean-Joseph saying he was celebrating the opening of the Dole rail line by sending his son a few cases of local white wine. An 1853 letter from Pasteur's wife, Marie Laurent, asks Jean-Joseph to send to Strasbourg, where Pasteur was professor of chemistry, a fifty-pound wheel of Gruyère because they eat so much of it—*"comme nous faisons une grande consommation."* In another request, from 1854, Pasteur is quite specific about the wines he requests—some half bottles of *vin de paille,* the famous local white wine that glows with a characteristic slightly oxidized flavor, and a case of red local wine (*"vins rouges très bons du pays"*).

There's something delightful about thinking of Pasteur opening a bottle of Poulsard at the end of a day, a grape variety from the Jura that we might see given a shout-out on a gastropub's chalkboard. Yet despite these occasional pleasures, Pasteur was a man who would let little take him away from his work. He was out to make a name for himself in French science—a small world comprising a handful of colleges and institutes around central Paris that existed in an atmosphere of patronage, blackballing for the Académie, and true excitement for the breakthroughs.

It was in this spirit that in 1852, when Pasteur was visiting Paris while still teaching at Strasbourg, the great crystallographer Jean Baptiste Biot sent a note to his hotel requesting Pasteur come to his lab-apartment at the Collège de France the next day to show his recently finished work on the shape of crystals to a German authority on the subject who was also in town. It is easy to imagine Pasteur at age twenty-nine rushing quickly across the cobbled courtyard of this famed institution tucked in behind a small square near the Sorbonne. What he was about to demonstrate was how crystals of identical shape in every facet imparted different properties to a beam of polarized light trained on their sides. Pasteur liked to explain this to his wife as two gloves that, though identical, cannot be superimposed because one is for the right hand and one for the left. He found that the light that didn't rotate came from inert materials such as quartz and reasoned that the light that did bounce back on a different axis reflected some form or quality inorganic crystals lacked. As French-born microbiologist René Dubos wrote, Pasteur's early experiment "created a new field of science—namely the relation of optical activity to molecular and crystalline structure." The shapes of crystals were a window into invisible life.

Though this is a long way from the bottle of malty *Hefeweizen* that boasts it is never pasteurized, the pieces that led to that technology are very neatly linked. In 1854, Pasteur was named professor of chemistry and dean of the newly opened faculty of sciences at the University of Lille in northern France. A certain Monsieur Bigo, father of one of his students and owner of a factory that used beet juice to produce technical-grade alcohol, approached Pasteur with a problem: many batches were souring before the juice could ferment fully. Using his polarimeter, a chamber for measuring the angle or rotation of polarized light, Pasteur observed that the crystals rotated; therefore, there was a living process at work. Turning to a microscope, he saw that in the vats of beet juice that soured, the round cells he recognized as yeast were outnumbered by even smaller oblong globules—what would eventually be identified as lactobacillus. He instructed Monsieur Bigo to toss all the contaminated juice, taught him to observe future samples to ensure the right bacteria had established itself, and, on August 3, 1857, read before the Société des Sciences de Lille a monograph on his observations that would be one of the defining documents of his career. The process of fermentation took many forms, but in each, the type of yeast was adapted to a particular food source. Because of cross-contamination, Monsieur Bigo had a confused fermentation on his hands. Instead of becoming alcohol, his juice was trying to become cheese.

Some years later, in September 1864, while back in Arbois on vacation he followed up on some earlier observations he'd made of local wines, setting up a basic lab in a café to study spoiled bottles the locals brought him. Cloudy and often oily, some wines tasted flat; others had an off-putting bitter flavor. Under the microscope Pasteur observed they were infected with *Mycoderma aceti,* a microorganism he was familiar with from earlier work on vinegar. The microorganism was inflicting low-grade damage on random vats throughout the area. Decades earlier, Nicolas Appert, the father of canning, had established that food cooked inside hermetically sealed containers did not spoil, so Pasteur knew that heating the liquid would help to preserve the wines. After experimenting with different temperatures—his son-in-law René Vallery-Radot credited the town tinsmith with supplying the braziers—Pasteur settled on between 50 and 60 degrees Celsius, high enough to destroy the microorganism without altering the taste. As was his way, a year later he formalized his findings in a paper he hoped would be useful to French industry. If the nation's wines were heated briefly, they could safely be transported throughout the world.

The technique's most widespread use was developed in 1886 by German chemist Franz von Soxhlet, who applied it to milk. It was a process that in the United States gained legitimacy a decade later when philanthropist Nathan Straus (co-owner of R. H. Macy and Abraham & Straus department stores) decided to have the milk he paid to have distributed to New York City's tenement children made safe. Whether or not it was false modesty, Pasteur alludes to the technique being named for him only in a footnote in his 1876 *Études sur la Bière,* a patriotic attempt to improve French beer after his homeland was defeated in the Franco-Prussian War. "They've even given the process the name of pasteurization," he writes, though one can see a hint of excitement peeking through the self-effacing placement of the comment.

AMERICAN BREWERS RECOGNIZED THE POTENTIAL for making their beers national brands and wasted no time in adopting the technology of pasteurization. Since European breweries had been slow to do so, there is a strong possibility that the American brewers tried the technique first, using information from the widely read technical journals. By the early 1870s, Adolphus Busch of St. Louis was using large bottle immersion pasteurizers and was quickly followed by the great breweries of Milwaukee, Cincinnati, and Brooklyn. This represented perhaps the most notable of several developments that transformed American brewing in a matter of decades. Techniques for the isolation of a pure yeast strain by Danish chemist E. C. Hansen at the Carlsberg laboratory in Copenhagen in the 1880s were immediately adopted by representatives of the Schlitz brewery in Milwaukee. That was soon followed by a mechanism that sealed bottles with crimped metal caps and finally left behind the wire-tightened corks that had previously been used. With the bottle cap, American beer could be enjoyed with a flick of a handheld opener or with a wall-mounted contraption the brewery salesmen handed out.

Now all the pieces were in place for big American brewers to produce tens of thousands of bottles and transport them across the country and remain confident that each would taste the same as what was being quaffed at the bar around the corner. The era of barrels was coming to the end and, with it, the job of the cooper, who had made them for centuries. It was laborious work. Staves that had aged outdoors for a year and a half—white oak was favored in Wisconsin—were heated, bent, and had dried cattail leaves placed between them to create the tightest fit before being gathered in metal hoops. A last necessary step for beer was to line the barrels with pitch extracted from evergreen trees, a final coating that was repeated each time the barrel was returned—via train and at the brewery's expense—from whatever distant saloon it had brought pleasure to.

The shift from wood kegs to metal bottle caps was accompanied by an incremental change in the size of breweries, which were transitioning also from businesses to major civic players. That community spirit is something the local microbrewery is emulating, whether by sponsoring a charity event, throwing open its tasting room for a fund-raiser, or gathering a few food trucks to raise money for a neighborhood cause as it gives tastes of its cask-aged ale. A hundred years ago community outreach might have involved loaning a team of Clydesdales to the Fourth of July parade or sponsoring a local choral society's trip to a distant World's Fair Exposition. It seems inordinately harsh to criticize the massive scale of these breweries particularly as the scale they brewed at is a feat unto itself. And the breweries thought of themselves as maintaining, perhaps even as the benevolent guardians of, their own very specific flavors. In 1890, an English syndicate saw the scale of American breweries as a ripe opportunity to integrate several of the largest into a single company profiting from every single economy of scale to dominate the nation's market. ("The advantage of world-wide advertisement incident to bringing the company out in London would be second only to Guinness," one typical inducement read.) Blatz Brewing in Milwaukee sold to the overseas investors but Captain Pabst didn't, and neither did Adolphus Busch in St. Louis. "Sixteen million dollars is enough to make a man's hair stand on end," Captain Pabst is reported to have written back then, "but I have been walking around this plant and I'm kind of proud of it. No, you can't have it."

YET DESPITE SUCH STRONG PERSONAL emotions by Pabst and others, big beer eventually came to be considered tasteless beer. All the energy that had once gone into solving technical questions was now focused on cutting expenses. The use of hops that added an elegant bitterness was reduced to trace amounts. The resulting bland suds created an opening for the plucky microbreweries with their sense of joy in rediscovering bygone flavors. They weren't, however, the opposites they might seem. Without big beer, small beer could not have happened. As a student at the University of Virginia in the 1970s, Charlie Papazian had become fascinated by brewing. A neighbor taught him and a bunch of his friends how to make beer from Pabst malt extract—a "dump and stir" recipe left over from home brewing during Prohibition. Later, after moving to Boulder, Colorado, he and his fellow home brewers depended on the goodwill of the brewers to get ingredients and basic materials. "In those days you bought malted barley by the tank car," the bearded pioneer whose 1984 book, *The Complete Joy of Home Brewing,* went on to encourage a generation of aficionados told me in an interview. "And a small brewery didn't need a tank car, so we'd drive to Coors right down the road in a pickup and get a far smaller load."

Probably no institution did more for the revival of brewing than the brew club. They existed in an atmosphere of learning where the jocular mood never eclipsed the self-imposed discipline of applying oneself to a craft. The nation's first such brew club, the Los Angeles–area Maltose Falcons, which began in 1974 in the back of the Home Beer, Wine, and Cheesemaking Shop in Woodland Hills, had a name with a pun—almost requisite in craft brewing—but also ran strict, informative tastings and classes. In addition, they maintained a yeast bank of rarities for those who might want to brew up a batch using Pilsner Urquell's yeast that a member with a little tech mojo had propagated from the legendary Czech beer.

Craft beer had a bit of a guerrilla warfare approach where ingenuity and smarts were always being pitted against the might of industrialized beer. Ken Grossman, an early member of the Maltose Falcons who went on to open the Sierra Nevada Brewing Company in Chico in 1980, defined the new spirit. The tech dreams that had been inspired in him as a child growing up in the San Fernando Valley reading books like *Tom Swift and His Megascope Space Prober* had progressed to customizing his bicycle as a teen. He eventually tried his hand at fermentation using Welch's grape juice and a packet of store-bought baking yeast. It didn't quite turn into wine, but it fizzed, bubbled, and made him want to know more.

Soon after graduating from high school, Grossman moved to Chico to be nearer the Sierra Nevada he loved to cycle in and hike. He opened a home-brew supply store and pulled shifts at a bicycle dealer; after meeting Paul Camusi, a fellow home brewer and cyclist, they decided they had the knowledge and at least some of the money to launch a small brewery. With accounts of these small, local breweries—Seattle's Redhook began in an old transmission shop in 1982—the technological thread that has run through brewing reached a poetic point. A new version of an old process was gaining a foothold in industries that had run their course. Grossman's account of cobbling together the equipment with which they started Sierra Nevada is nothing less than the recollections of a gearhead let loose. The compressor and coil for the refrigeration came out of the walk-in cooler from a neighborhood butcher shop that had been shut down by the health department. The old milk tanks from the small dairies in Northern California served as fermenting vats. The longnecks were from the Maier Brewing Company of L.A. shuttered in 1972. The bottling lines they ran on came from local soda bottlers. As Grossman writes in the company's history, *Beyond the Pale,* "I had no operation or parts manual, so I just had to figure out how it was supposed to work. Luckily we still had two small soft drink bottling companies operating in Chico at the time, and I started picking their brains regularly. The very experienced foreman at the tiny 7UP bottling plant was extremely helpful, having operated similar equipment. For the first year of operation, both of the local bottlers would regularly let me have their almost empty glue buckets to scrape the bottom for the few cups of label glue that we required to label our 100-case runs." All that tinkering that he loved to do now found focus in fitting the salvaged equipment into a three-thousand-square-foot warehouse in an industrial street south of town.

GO TO ANY BEER BAR today and between what they have on tap and in the cooler you can pretty much have anything from the wide world of brewing. There are fresh farmhouse ales, enticingly fruity lambics; if it's hot outside, you can cool yourself with the light, straw-colored kölsch they favor in the German town of Cologne. There are dark stouts, too, and red-hued monastic beers that climb into downy soft foamy heads when poured. If they have a beer on special, more than likely you'll be told about the barrels it might have fermented in or the touch of rye malt that contributes to the complexity. The abundance of information doesn't help, though, with defining what craft brewing actually is. With big beer, that was easy. At every point, it sacrificed characteristics for yield. Craft brewing is not so clear-cut. The big differentiator is considered size, and there's endless disagreement about the production level at which one is no longer a craft brewer. The Boulder-based Brewers Association first said it was more than two million barrels a year. When Samuel Adams protested in 2010, the figure was adjusted to more than six million. A lot of it is intended to call out the megabrewers who have created boutique brands that make them appear small. But does it really matter? Craft isn't measured in volume but in proficiency and intentionality. Having a cutoff number might be trying to measure what cannot be. Who's to say that the brewer going to the industrial-plant-sized brewery is not approaching it in the right spirit, or that the beer only available on consignment isn't compromised by its exclusivity?

What the craft movement did was approach brewing with a new spirit. It found an irrepressible joy in processes and steps that had been rendered obsolete. In an era when food and drink were being made on an epic scale, resourcefulness meant everything, and if the hidden-away store smelling of malted barley where the Maltose Falcons still meet (first Sunday of the month) has a much bigger selection of specialized yeasts than it used to in the early days, the Falcons have not lost their tenacity or the sense that anyone who applies themselves can learn to make good beer.

In the back room on a recent Sunday, a forty-strong group of mostly men in all forms of cutoffs and flannel shirts sit in plastic chairs. They trade tips on decoction and fishing spots until one of the club's officers sounds a gavel to call everyone to order. The club's Grand Hydrometer, a man with a trimmed Vandyke and an easy smile, rises from behind a plastic foldout table to talk about seasonal beers. People are still talking. "Pipe down," says a young man who, wearing jeans tucked into rubber boots, has just rinsed out the club's brewing kettle. It's a day of learning and tasting and plans for visiting breweries in other cities. After lunch—porter-braised beef stew—members introduce the beers they've brewed and brought in, detailing their malt mixes, yeast types, and the like, often reading the recipes from the screen of their phones. One by one samples are poured into plastic cups. Knowledge is displayed. *Burtonized* is used as an adjective. One gray-haired guy in a Munich tourist T-shirt mentions phenols, nuances that tend toward clove and spices. When another guy who's known for being vocal says he can barely detect the sage in a farmhouse braggot, a meadlike beverage made from malt and honey that lends itself to a variety of flavorings, someone else gets a chuckle out of him, playing off the word and quipping, "That's because you are one."

The question is: Why go to all the trouble? If you imagine the world outside this crowded room, you can see shelf upon shelf in liquor stores and supermarkets filled with every type of beer. But on the shelf, good as they are, those beers are product; in the hoppy confines of this tight space, beer is a process. Okay, and maybe an obsession, too. “The whole point of the hobby,” says one guy in tube socks when I ask why he joined, “is to learn.” Perhaps that’s the simple retort to industrial efficiency. Craft was reinvented, not as lore but as something living and vital, something weekend dabblers who find joy in each step and task can experience by making a recipe repeatedly and seeing how their skills improve with each one.

PART THREE

TO ESSENCE

I'm at the wheel of a rental car heading across southwestern Wisconsin en route to Uplands Cheese, located outside the town of Dodgeville. I pass mostly farms with their concrete silos and windbreaks. It's early spring and a thin coating of snow clings to the corn stubble in the open fields.

Uplands is owned by two couples: Scott and Liana Mericka and Andy and Caitlin Hatch. Scott is in charge of the farm's 150 cows; Andy makes two cheeses from their milk. Pleasant Ridge Reserve, a dense, Alpine-style ten-pound wheel with a firm interior and a natural yellowish-brown exterior that, like the best Swiss cheeses, grows ever more intense as it ages and develops a natural rind, is the original of the two. It is made only from May through October, when the cows eat lush grass. The other one is Rush Creek Reserve. Runnier than Pleasant Ridge Reserve and intended to be eaten fresh, it is modeled on Vacherin Mont d'Or, a storied mountain cheese that's made both in France's Franche-Comté region and across the craggy border in the Vaud canton of Switzerland—in autumn, when the cows are fed hay. Each small disk is wrapped in a spruce bark band that keeps the creamy interior from flowing out and contributes a little nuance, too, like what your hand smells like for a few seconds after you toss a log on a fire. With Vacherin, the aroma lingers on the surface of the nubby puck, making it even more complex.

It's the firmer Pleasant Ridge Reserve that I have come to find out more about. I've become fascinated by this cheese wheel with its smooth rind that feels like sandstone and covers a golden-yellow interior. I can't cut a slice without wanting to bury my nose in it. The taste: buttery, toasty, nutty . . . whenever I try to define it, my mind shifts from adjectives to verbs. The cheese hints at something fresh and vital, while retaining the glow of an ingredient that's been burnished, rendered to its essence. Like pasture, it's still a little wild and untended, but those aspects have been transformed into something fine.

Food description is a contradiction in my line of work. Ostensibly, it's one of the main reasons someone would read a review, but too much information and you quickly lose the reader as dish description starts to sound like shopping lists. I've seen words appear in the vernacular, glow with new meaning, and quickly become shopworn. Can every ingredient be heritage, every piece of equipment vintage, and every microbe native? Over time, certain truths have become self-evident. *Bourbon* and a hyphen make anything sound delicious, *sugar* is a term so studiously avoided you'd think no one ever uses the ingredient itself, and *slow* represents integrity, especially when tacked on to *roasted* or *braised*.

The fluidity of language has given me a heightened interest in how the great practitioners have negotiated the task of evoking taste. Joseph Mitchell used mere facts brilliantly; somehow by grouping them together in a certain way, he gave them an associative power they'd never have alone. The butcher preparing for a beefsteak dinner, a Manhattan version of a riverbank fish-fry, "had carved steaks off thirty-five steer shells and had cut up four hundred and fifty double-rib lamb chops," he wrote in "All You Can Hold for Five Bucks," his 1939 *New Yorker* piece on the rituals of these club get-togethers. "In his icebox, four hundred and fifty lamb kidneys were soaking in a wooden tub." Hardly heavy on adjectives, yet you sense the cheer that these cuts would bring when seared and consumed in uptown halls with brassy German bands.

Geographical specificity is another way of evoking flavor. Every day I read menus that list place names, narrowing the world to a hollow or a farm as if inviting me to make the mental journey to a rural paradise. Drawing a bead on a locale today is intended in a different way than it was decades ago, when a sign in a grocery store window let shoppers know the Jersey beefsteak tomatoes or Texas onions had arrived, each with a distinct quality of ripe sweetness. It's also not applied as it is in the domain of, say, single malt Scotch, where a whisky from the island of Islay radiates a special peaty quality while a Speyside is milder. Our version of provenance is almost allegorical—it's intended to communicate two things at once. The first is simply the location where the food is made, but locale is also intended to evoke all sorts of positive connotations. In the 1930s, when the first metal cans in the shape of log cabins started to come out selling Vermont maple syrup, the state's name was merely a helpful way of communicating the syrup wasn't from New Hampshire, Quebec, or Maine. Today, the word *Vermont* on any label casts a glow of something unsullied, uncorrupted, almost spiritually whole.

Perhaps it's because food has become so estranged from an actual provenance that invoking the origin point heightens the meaning. Still, when the evocative power is abused and there's a pileup of farm names, breeds, and old-time processes, I sometimes think the idea of indigenous flavor is being burdened with re-creating the bonds that industrialization destroyed. That's a daunting responsibility, but I can't deny that great flavors, true and clear at their center and gleaming with slowly revealed facets, are capable of the feat. I've been stopped in my tracks by one bite of Benton's bacon (from Madisonville, Tennessee) tucked into a high-stacked BLT. Now I'm excited to see one of those places a unique food comes from. In the center of Dodgeville I take a sharp turn north onto State Road 23. The road dips and rises into open land. Within minutes, I'm pulling up in front of a white farmhouse, a creamery with an office, and sheds and various pieces of equipment arrayed behind. The Wisconsin landscape unfurls toward the horizon, flat and brown under a patchwork of snow.

ANDY HATCH IS THIRTY-FIVE YEARS OLD. He wears Dickies flat-front twill workpants and a brown plaid shirt under a smock that resembles a chef's jacket and a traditional cheesemaker's cap, a short-visored covering that serves to control hair and proudly announce profession. He owns a large collection of them, which he keeps in the plant's changing room. Over a welcoming slice of good, heavy poppy seed cake and coffee, we sit in the small office weaving in and out of subjects such as the revitalization of downtown L.A. We're just getting acquainted. His wife, Caitlin, designs and paints a new promotional poster for the farm every year. Liana, Scott's wife, bottle-feeds the calves in the spring. To relax, Andy plays mandolin in a bluegrass band called Point Five, an allusion to Mineral Point, the town where all five members met. They perform at a variety of local venues including breweries, free-clinic fund-raisers, and the county fair. Andy also likes playing in his hometown opera, a onetime vaudeville house with scalloped balconies and a small stage with good acoustics. "It's nice," he says of membership in the group. "The doctor who delivered our kids plays acoustic guitar."

As with many artisans, there is a sense that his story starts with his discovery of craft in general. At a certain point, a casual interest becomes an identity—as mind, motivation, and trajectory narrow to the single purpose of becoming an expert at one sole activity. Often, though, it's the activity that seems to single out the person. The son of a Milwaukee lawyer, Andy's artisan story started soon after graduating from Trinity College in Connecticut when he enrolled in the Michael Fields Agricultural Institute, a biodynamic research organization where he intended to study corn breeding under Walter Goldstein, a noted corn breeder and agronomist. It wasn't yet a craft, but he'd chosen the direction it would branch out from: agriculture.

Though Goldstein describes biodynamic practices holistically as "a view of life having a reality to it above the modern chemical and physical conception," as a scientist he approached life concretely. Corn, he was given to tell the volunteers and students who gathered at the center, had become "genetic machinery," owned by two or three companies. Through plant breeding—not genetic modification—and working from heirloom corn he'd sourced himself from a town in northern Nebraska, he sought to breed new varieties of these corns that are protein-rich and delicious, better adapted to their climate, capable of living from (and contributing to) organic matter in the soil and of growing in open-pollinated fields. A tall order in an age of patented seed and Roundup weed killer. But from the start, Goldstein had been encouraged by farmers, who saw him as a visionary who could reestablish age-old qualities in the plant without sacrificing yield. In other words, he was developing an ancient—yet modern and viable—crop.

There was something tangible to the work that Andy responded to once he'd started. He would walk through the fields amid the stalks (many kept wrapped in newspapers to prevent contamination of genetic traits), encouraged by the real-world applications of what he and his colleagues were doing, such as organic poultry feed and tasty, non-GMO corn products within the large business of commercial food production. In contrast to manufacturing methods where flavoring agents would mask any deficit of taste, they considered the corn's real flavor a definite advantage, and naturally one of the parameters of quality in the extensive trials was flavor. "After harvest we'd work our way through a windowless room where twenty thousand ears of corn were stored," Andy recalls. "We'd make little muffins from the corn with water, baking soda, and very precise amounts of salt. The variety of flavors was huge. One corn might be perfumey, another lovely and have a natural butteriness." Still, Andy was also starting to realize that the project involved more lab work than he'd imagined. Andy is compact and thin and, as he sips a cup of coffee, capable of projecting an intense seriousness that gives way to an easy smile. He's not going to come up with some complicated reason for why he left. "I was twenty-four years old," he says. "What I really wanted to do was travel."

Goldstein's Norwegian wife, Bente, had recently lost her father, and the couple suggested that Andy go to Norway and help Bente's mother, Unni, with all the extra work she now faced on her farm. To Andy, this seemed like a great idea. He could continue his search to find some form of agricultural profession that was more suited to his interests and abilities and see a new place at the same time.

Unni kept things simple. Her land, about halfway up the Norwegian coast near Molde, climbed the steep slope of a fjord at such an angle that all grass-cutting had to be done with a scythe. Hay was dried in the traditional manner, draped over a series of steel clotheslines. The goats were milked by hand in a hut that looked out toward an expanse of crystalline ocean.

The young American and the elderly Norwegian lady formed an easy friendship, and as winter approached she encouraged him to continue his travels. Keeping to a network of biodynamic farms, he spent time with an Austrian family who'd had their land for eight generations and followed that with a stint in Italy. He spent one year in western Ireland living with a couple who kept sheep and widened his musical experience with a standing Sunday night gig at the Royal Spa Hotel in Lisdoonvarna, where Andy played the mandolin and the farmer played the handheld *bodhran* drum. He had moved on to Germany in 2006 when he received a phone call. "It was the sort of phone call you see in movies," he says. "Your father is very sick. Rush home."

Before Andy can continue with the story, a van pulls up outside, and the man from Ecolab, a sanitation service, walks in for the monthly visit. It's not the usual technician, who's apparently home with a twisted knee. "He told me he wishes every place was like you guys," he says as Andy gets up to greet him. After some back-and-forth with a clipboard, Andy hands him booties and a hairnet and they go inside.

LEFT ALONE, I GET A chance to look around the small office. I see the expected computer, printer, calendar lining one side of the room, a bookshelf along the other. Among the books, I notice Dominik Flammer's tome *Swiss Cheese,* a celebration of the craggy peaks and lush valleys that create the massive traditional wheels that are the distant models for Pleasant Ridge Reserve.

It's a full-color tome that covers more than a thousand years, going back to the earliest days of Swiss cheesemaking when the milk (left over after the fat had been skimmed for butter) was soured and pressed into blocks. Without rennet, the cheese never passed through the curd stage; instead they were crumbly, tangy, and very fresh—squares of Bloderkäse still found in the region of St. Gallen speak to that early farmhouse tradition—but they are anachronisms in a land that by the fifteenth century was already exporting the big wheels we are familiar with through Alpine passes. Writing on the origins of that immediately recognizable shape, Paul Kindstedt of the University of Vermont points to topography and distance to market as key factors. A soft ripened cheese such as Brie could be flat and runny because there wasn't a great distance to travel from Meaux to the markets of Paris. However, the roots of the Alpine cheese family derive from communal herds that feed in mountain heights during spring and summer. The large volume of milk they produce needed to be transformed into big wheels of cheese and given a sturdy enough natural rind to withstand being packed back down to the valley floor at season's end. Voilà le Gruyère.

Though Pleasant Ridge Reserve shares a sturdy rind with (and inspiration from!) those storied cheeses, its roots are devoid of any of that kind of folklore. In 1994, the original owners, Mike Gingrich and Dan Patenaude, together with their wives, Carol and Jeanne, bought the farm from a family who'd worked it for thirty years. Though the southwestern Wisconsin landscape of silos and barns suggests long links to the land, this isn't particularly arable soil, and farmers have been negotiating it and trying to optimize it for many decades. One previous owner planted a stand of walnut trees in the steepest section when tractors replaced horses for plowing, figuring at least he'd harvest something there. The family the Gingriches and Patenaudes bought it from kept a small herd of cows in a stanchion barn, a structure that keeps the animals in one place as opposed to letting them roam. It's a practical system because the animals gain weight quickly and are always there at milking time. There's also a lot of manure to deal with. On the land itself, they grew alfalfa and corn. "They'd already raised their kids," Mike recalled when I called him, "and they were at an age where they didn't want to keep keeping cows."

The new owners saw it differently, almost diametrically so. Inspired by a book on rotational grazing that Jeanne's brother, Bill Murphy, had written, they were going to give the land over to the hustle for resources that mixed foraging represents. Published in 1987, Murphy's book, *Greener Pastures on Your Side of the Fence,* paints pastureland as something quite different from the peaceable loam one might romanticize about. The clover, the quack grass, the legumes, and the secondary grasses are in a constant state of competition for nutrients, water, and, since leaves cast shadows, even for light. Murphy credits grazing cattle as preventing the grasses with dominant traits from rendering the landscape as uniform as a suburban lawn.

"Animals always cause a pasture to be a more complex mix of plants than it otherwise would be," he writes. "This is because animals graze selectively and in patches, and the effects vary in time and in space. Animals also drop their manure and urine in patches, which affects some plants more than others. Besides walking, running, and jumping on a pasture, animals sit, lie, scratch, and paw on it. All of these things result in a sward containing a wide variety of plants adapted enough to survive their different local conditions."

The plan was that Dan, an experienced herdsman, would institute the system of movable paddocks that would encourage lush varied grasses. Mike would use the primo milk they derived from it to make cheese. Several hurdles appeared immediately. The first was that they had no idea what kind of cheese they wanted to make. Second, Mike, who had worked mainly as a manager at Xerox in California, did not actually know how to make cheese.

It was Mike's nostalgic memories of family visits to midwestern farms as a child that initially made him want to move to the region and farm. Over time, the endearingly blunt ways of the country had stuck to him. It's not just the wire-rimmed glasses and the triangle of white T-shirt framed above whatever he's wearing but also the way he speaks. When we spoke on the phone, he'd become very specific about why the grass is best eaten at the moment, "before it heads out, when all the energy goes into the formation of the seed." He didn't deal in superlatives—few serious artisans do—and he'd been matter-of-fact about how they'd solved their two most immediate problems.

"I had samples sent from Murray's that I had identified from the Steve Jenkins book," Mike recalled, "and a group of us sat around the kitchen deciding which one we liked." Murray's is Rob Kaufelt's famed Greenwich Village cheese store, and Jenkins is the celebrated cheese expert of the Fairway Markets throughout New York. While the wealth of knowledge Gingrich and Patenaude were drawing from is impressive, there is something wonderfully insouciant about deciding what you're about to devote your life to based on what goes well with crackers and a crisp chardonnay. What they chose to model their cheese on was Beaufort, an Alpine-style wheel from Savoie, just across the Alps from Switzerland in eastern France. Now Mike just had to learn how to make it. Luckily, he was in the right state.

What the University of California at Davis is to wine the University of Wisconsin–Madison is to cheese. The department of food science, which has been part of the College of Agricultural and Life Sciences for more than one hundred years, has all sorts of bragging rights. Preeminent among them is the 1890 invention by agricultural chemist Stephen Moulton Babcock of the centrifuge-like machine, called the Babcock Milk Tester, capable of testing the butterfat content in milk. The machine was simple enough that every farm soon had one after its development. Knowing the fat content immediately established a basis for price and created an incentive to improve herd stock, as well as a strong disincentive to those who might skim or water down the milk. Since then the department has been a leader in dairy practices and, while remaining focused on commercial cheesemaking, evolved to guide the growing numbers of small-scale cheesemakers with the launch of the Center for Dairy Research in 1986. Mike enrolled in the Cheese Tech Short Course. The one-week courses originally ran from November to April when the sons of farmers could get some time away from the farm. Today, you can sign up for year-round courses that run several months, or for shorter, more intensive workshops focused on cheese grading or commercial ice cream manufacture, but each reflects how the university has been an active partner to the state's dairy farms as their needs shift.

Mike was one of the small-scale producers whose needs the

university had started to address. After taking the course, which helped him gain the necessary certification, life became centered around Babcock Hall, the midcentury brick building home to the dairy science department. Once a month he brought in milk from the farm and set about devising the "make procedure," or recipe, with department scientists John Jaeggi, Mark Johnson, and Jim Path. These were food academics with a passion for cheese—Path had a side project of introducing to America a Finnish baked curd cheese called Juustoleipa—and over a two-year period they introduced Mike to the rigors of developing a new cheese. There were different combinations of microbial cultures that would acidify the milk and leave a lasting imprint as the cheese matured, different ways to develop a rind. It would be months before they could even taste the results.

"Mike would never stray from detail," Jaeggi recalled when I called him. "Cheese will go through stages, and every time we tasted, he'd make careful notes. Is it too acid, too bitter, how's the surface ripening?" Jaeggi was impressed by Mike's focus and how open he was to instruction. The beginner was determined to make a great cheese, and two years later, when the recipe had finally been set, he came to an arrangement with Bob Wills, a pioneer in hormone-free cheesemaking at the Cedar Grove plant in Plain, a small town on the other side of the Wisconsin River from Dodgeville, to make it there. Jaeggi went up to Plain on a Saturday and helped Gingrich make the first batch.

A deal was struck—Mike could make his cheeses there on weekends as long as the vats were clean and the cheeses were off the presses before dawn on Monday mornings when Wills needed them. Cleaning cheese equipment is no casual thing and takes hours. It required Mike to start cleaning by 3 A.M.

Mike still recalls those dawns when he'd be bundling his freshly formed wheels into clear plastic bags, placing them into the back of his Ford Explorer, and driving back to the small town of Spring Green where he'd rented a walk-in cooler from a small food producer. This was his "smear room," the place where the wheels were rubbed in salt and put on wood boards to begin the process of maturation. In 2000, the Gingriches and Patenaudes first sold their cheese to the public. Eager to see how it would do against other cheeses, they also entered it in the American Cheese Society competition being held in 2001 in Louisville, Kentucky. It won first place in its category and best of show overall.

ANDY AND THE ECOLAB TECHNICIAN reappear and there's some more form signing on the clipboard. The man eases off his booties. "Put a lot of those on?" I ask as he tugs one off the heel of his shoe. "This is nothing," he says, "you should see what I have to wear in a meat plant." With that he stands up and tears a record of the visit off the clipboard. Andy opens a metal cabinet and files it immediately. "Where was I?" he says, sitting back down.

"Your dad," I answer.

In 2006, back home and with his father's health improving, he understood that his travels had initiated and eventually confirmed his wish to be a cheesemaker. "People think you get handed some kind of golden manual when you go to Europe," Andy said with a laugh. He'd heard of Pleasant Ridge, and the rotational grazing method used there particularly appealed to the environmental aspect of his interest in farming. Soon after returning from Germany, he drove out to Uplands to ask Mike for a job. It was too early, the older man said. He needed to learn much more before he could begin.

And so, just as Mike had, Andy enrolled in the Cheese Tech Short Course and, additionally, the Farming Short Course. He was twenty-six years old. In addition to gaining knowledge in the basic dairy science procedures, he was intent on using his time at Madison to thoroughly learn to make cheese. Again the university came through, this time in the person of Gary Grossen, a Gouda expert who transformed the milk from the department's eighty-five-head herd into a variety of cheeses used in student dining and sold in the sampler boxes and attractive special gift baskets in the Babcock Hall store. (The milk also is used for a robust ice cream program, and it's not rare to see someone crossing the quad nestling a pint of Cinnamon Snickerdoodle.)

The artisan story is sometimes told as one where newer cheesemakers rejected cheese made by big commercial outfits who'd chosen volume over nuance. As marketing it's effective; however, it's not accurate. The big blocks of sharp Colby, lacy, sliced Swiss, and chive-flecked Monterey Jack that fill the nation's dairy cases may not have evocative packaging, but the best of them affirm scrupulously executed craft, all the more inspiring because the cheeses are destined for kitchen counter snacks, lunch pails, and brown bag sandwiches. An affable, avuncular type, Grossen is a direct link to that older Wisconsin dairy tradition. To one and all—and certainly a journalist calling him for an interview—he insists they visit the National Historic Cheesemaking Center and Museum in his hometown of Monroe, the seat of the state's Swiss cheesemaking tradition. He grew up there, above the plant his father managed, making the boxes for shipping and always alert to what his dad and other workers were doing. Those were hardly the days of big tasteless wheels. "We'd make our own culture from cooked whey," he recalled to me in a proud voice, leaving the implication hanging that even the strictest artisan today uses commercially produced cultures for the initial souring of the milk.

What Grossen did for Andy was slow cheesemaking down. The wonder of cheesemaking is that it starts so simply yet transforms into something so complex. There are processes of acidification to control the milk, certain enzymes to separate protein from liquid, and then steps like deciding the size the curd should be cut into (layered mats of small curd, for example, will lead to cheddar). From a lone liquid, you somehow end up with objects as different as a nugget of crumbly Parmigiano-Reggiano and a green-veined wrinkly orb of New Zealand Cloudy Mountain cheese. To get to the complex, you have to take command of the simple, and that is what Grossen concentrated on. Peppered with little truisms, like "there's a little Swiss in all of us," he laid down the parameters: have a system, don't be afraid to ask for help, keep records. A lot of it was hard-earned touch; some of it was trained eyesight. Grossen could eyeball a vat and know precisely the moment when it would cut cleanly and not shatter. "He'd make one vat a day and at first I'd just watch him. He took the time to explain every single thing to me," Andy says as we finish our coffee and poppy seed cake. With Grossen's expert teaching as a foundation, Andy returned to Uplands when he graduated, and Mike took him on.

I FIRST TASTED THE PLEASANT Ridge Reserve in 2008—seven years after its breakout award—at the American Cheese Society conference held at the Chicago Hilton, a grand building on Michigan Avenue across from Grant Park. Long gone are the days when the society met in places like a modest motel in Bird-in-Hand, Pennsylvania. This hotel, with its marble hallways and sweeping staircases, had a baroque magnificence. However, as per my usual experience, the convention area was more utilitarian—durable carpet, soundproof tiled ceilings, charging stations. I'd exchanged cards with someone from the Jersey Cow Association, sat in on a seminar about the chemical properties of butter, and listened in on a panel where a representative of a large national market had referred to the cheeses it carried as an SKU, grocery biz acronym for *stock keeping unit,* an abrupt reminder of the barcode-operated world outside.

If I was ready for a little cheese lyricism—that realm where the pert Nubian goats are always cocking their heads with great interspecies curiosity—I found it that night at the grand tasting—always an anticipated part of the gathering. The hotel's massive gilded ballroom was filled with long tables loaded with cut samples, which the individual cheesemaker was often helping to serve. There were bloomy rind wheels, cut at just the right temperature to show off the creaminess they contained, and cheddars that had gained a sharp sandy quality from patient aging. One blue, from Rogue Creamery, wrapped in grape leaves macerated in pear brandy from Clear Creek Distillery, was a lush standout, the diamond-cut intensity of the spirit seeming like the only taste capable of framing the cheese's force.

Even with all this competition, the Pleasant Ridge was still memorable. Mike was standing behind a table with Andy, who by then had been working for him for a year. They were just one more duo cutting and putting out samples in the milling crowd, though there was something about them—perhaps it was the unmatching plaid shirts—that made them seem consciously connected to a farming way of life as they stood side by side.

In 2007, when Andy had returned to Dodgeville looking for

a job, Mike saw in him a young man who just might be able to push the cheese forward into a newer sphere. Andy inhabited that realm. Soon after starting, he was already encouraging Mike to age the wheels longer to see what nuances might develop. That might sound like one is pushing flavor the way a track coach with a stopwatch drives a runner, but it's more like a theater impresario who gathers a great cast of bacteria, yeasts, molds, and enzymes and ensures they have time and room to do their thing. Andy shies from including his own skill in the process. When I ask him what it is that makes his cheese unique, he doesn't opt for a lyrical answer. He doesn't even talk about craft. Instead he gives me a list of concrete facts. "Our cows, our pasture, raw milk that's never more than twelve hours old, careful ripening, and a natural rind."

The great milk is Scott's responsibility. An affable thirty-year-old with a scraggly beard, Scott is originally from North Carolina, where he grew up on a corn and tobacco farm and studied agronomy at Warren Wilson College in Asheville. Scott got a job at Uplands in 2010, and soon proved himself a natural with the cattle—and with the farm's core component: the grass. Dodgeville is located in what's known as the Driftless, a massive area reaching from Madison to the Mississippi and across the river into Iowa that dodged a glacial period, which resulted in the high elevations called uplands that gave the farm its name. Though it has pretty dips and hollows, there are no Alpine heights here. By constantly moving the cows, a well-disposed cross of New Zealand Jersey and New Zealand Friesian, to pens enclosing tall fresh grass, Scott recreates the effects of a herd of brown bell-ringing Swiss cows mowing their way up the slope of a mountain to ever-higher pasture.

Hardy creatures, the cows remain outside all year. In winter, Scott takes hay to the cows where they shelter in windbreaks.

"They're like a wood-burning stove," he says of their ability to remain warm. "You just have to keep them fed." During spring and summer, Scott is up in the fields at five A.M. with his ATV and his collie, Brier. She is gentle with the cows, suggesting with a flick of her tail or a nod the direction they should take out of the nine- to eleven-acre paddock where they've grazed. Not that they really need it. The cows were selected for their fertility, their easy dispositions, and their ability to thrive on grass, and they know the drill. "If we're not there on time, they're looking at us like, 'What took you so long?'" The cows immediately start ambling down the lush raceways to the milking parlor where they stand side by side in rows of eighteen, resembling distracted passengers waiting to board a flight. This milk will be added to the milking from the previous evening and will flow, raw, thick with butterfat, a palest shade of a buttercup yellow, into the vat where it will be transformed into cheese.

NOW WE'RE STANDING AT THAT vat, toward the end of the tour Andy has given me. It's empty, but I'm still wearing a hairnet and a shop coat as required. *Raw,* as it's used in cheesemaking, means unpasteurized. I've used the word many times, but now it catches in my ear. It's a curious word to apply to milk. In his 1978 book, *First Person Rural,* Noel Perrin, one of the more literary of the era's back-to-the-landers (he was an English professor at Dartmouth College when not writing odes to farmwork), found himself questioning it as a term. "Raw is what uncooked meat is," he wrote, "with the strong implication it *ought* to be cooked. And so it should. But milk is a finished product as it comes from the cow. One might as well speak of raw orange juice, or a breast-fed baby guzzling raw mother's milk."

He was prescient about the word, at least if you go by the number of cold-pressed juiceries in L.A. who've latched on to *raw*'s cleansing connotations. The root of the new usage, though, dates to a series of studies conducted in the 1930s and 1940s by physician Francis M. Pottenger Jr. who contrasted the health of cats fed an unprocessed diet—including "raw milk"—with that of cats whose food was cooked and found it to be far superior. Today the term has taken on suggestions of a more harmonious relationship with the environment than heat-treated milk can provide. The way Andy uses it is practical; this liquid is whole and unmodified, buzzing with the forty to seventy species of microbes French cheesemaking consultant Ivan Larcher estimates thrive in high-quality raw milk, each one of which plays a role in the complexity of the finished cheese.

The milk hits the vat in the morning at about 75 degrees Fahrenheit. It's a mixture of the previous evening's milk, which has been stored overnight in a tank at 40 degrees, and that morning's cow temp milk. All Andy does at this stage is raise the temperature ever so slightly to 90 degrees. What he is in fact doing is giving the microbes native to the farm's raw milk the opportunity to grow, the first of several steps where he allows naturally occurring processes to get going before he enters the equation. He adds a tiny amount of acidifying culture after an hour, a period that has allowed the milk's complex mix of microbes to flourish so that by the time the ones he's added have reached critical mass, there's a fair competition for the milk's lactose. "By delaying that moment, there are no dominating cultures," he says. The pauses he's building into the procedure will all be reflected months away in nuances in the cheese. It's only because he has complete faith in the quality of his milk that he can afford to wait.

After the rennet is added, the liquid will become gelatinous as it coagulates, and, after cutting, the curd temperature will be raised to between 90 and 120 degrees Fahrenheit as it spends several hours cooking in the vat. This process replicates what happens in those massive copper kettles used in the Alps, but Andy prolongs the stage for a longer period than most Alpine cheese recipes call for. His approach, always alert to encouraging nuance, is "weighted toward complexity," and he considers that extra time a necessary step. Eventually, the vat is a twelve-foot-by-four-foot squiggly block. What's left is to drain the whey and then pack the curds into food-grade plastic molds. These get stacked on top of one another in a slanted press—so the remaining whey doesn't drain from wheels onto the wheels below—and Andy uses two hands to pull on the levered crank, applying pressure that will fix the cheese's form.

When released the next morning, the wheels are a raw paste—the most basic form of cheese, a coagulated mass solid enough that it can be picked up. Each one is inert, silent—pure potential. Andy can be funny and offhand about this stage. "What is cheese but a big lump of fat and protein," he says as we enter one of the aging rooms. It's humid, filled with racks bearing the wooden boards each wheel rests on. It smells of the ammonia they release as they age. This is where what's latent becomes realized, where that big lump becomes something unique. "Using raw milk leaves in all the enzymes that will metabolize the fats and proteins," he says, walking over to one of the racks.

The cheeses on this rack are four months old. The date is written with a Sharpie on a plastic card tied on the rack's edge. The color of the cheese is starting to turn from a light, butterfat yellow but hasn't yet reached the sepia brown of the older wheels. The rind is the result of the repeated rubbing with salt, each application drawing moisture to the surface that will interact with the salt-tolerant yeasts that are busy there and lay down another microscopic layer. That's different from the cloth-covered drum-shaped truckles of English cheddar and the foil-wrapped pitted surface of Roquefort. It's not downy soft like a bloomy rind wheel, or crinkly like a puck wrapped in chestnut leaves, or achingly pure like a goat cheese pyramid dusted with ash. Every cheese rind is the result of an interaction between nature and environment, as well as an identity.

Andy takes a metal cheese trier and pushes it into the side of a wheel. One wheel per rack is sacrificed for tasting purposes, and the hole is carefully plugged afterward. Depending on his opinion, it goes to market between eight months and two years. I take a small piece off the metal taster and rub it into a ball between thumb and forefinger the way he's taught me to, warming it and releasing its flavors. It tastes fresh; there's a hint of grass there, but its faintness is more like the memory of a field. I'm reminded of what Virginia Woolf said when writing to thank Vita Sackville-West for butter during the rationing of World War II. "You've forgotten what butter tastes like," she wrote. "So I'll tell you—it's something between dew and honey." There is something both unsullied and unalloyed there, and it is magnified when we taste a piece of a cheese that's a year older. The first cheese retains a coltish freshness; the second is turning toward golden flavor—with a pronounced sharpness.

These nuances remind me of how each wheel represents a willed compression. It starts with the grasses and wildflowers growing in each of the ten-acre paddocks. That herbal variety is translated into eight thousand pounds of milk, which becomes eight hundred pounds of cheese, which, by aging and losing moisture on a rack in this room, becomes an essential reflection of a time and a place. We end with a two-year-old wheel, and I get to understand in a new way what time does. Great artisanal foods often have a quality where whatever might have been extraneous has been consumed in the process. That's how it is with great wine vinegar or bourbon, products that have aged long enough to curb all the raw notes. This cheese is at that stage when I taste it. The tang is gone, the notes are resolved. It is like glimpsing what all the beautiful grass has become. "I sometimes think that there is a form of return," Andy says as he tastes it. "There's a sweetness in the youngest cheese and a sweetness in the older ones. It's as if it comes back."

"Let me cut you some for your trip," he goes on as we walk back out of the aging rooms. In fact, a piece has already been cut for me and wrapped with great care in waxed paper with the rustic Pleasant Ridge Reserve label as seal. It is a big wedge of the cheese I now know something about. It feels very real in my hand as I take my leave and go back outside to my rental car. As I start down the incline of State Road 23, the benchland opens up and farms with their belts of trees reappear and I feel something powerful.

I like to think that history adds a nuance to my wedge of cheese. It may not be the result of handed-down tradition as in Europe, but it has been colored by an American passion. I think of Andy playing the mandolin in a repurposed country opera house and Scott heading up the tracks in the morning as reestablishing a rural ideal. Those things are reflected in the cheese that I have with me when I return the rental car, and when I explain it to the curious TSA agent, and finally, when I cut into it with a plastic food services knife during a long layover in Dallas/Fort Worth Airport.

TO WORK

Mike Conlon was nursing a shot of bourbon as only a man who has spent all night barbecuing outdoors in Brooklyn in February might. He was in no rush. He'd pulled forty briskets, ten pork shoulders, and sixty racks of beef ribs out of black steel smokers a few hours earlier. Together with his boss, Billy Durney, he had also loaded it all into insulated food storage Cambro containers and driven the barbecued meat over the few blocks from the open space where it had cooked to the restaurant where it would be served.

Durney is a new breed of pitmaster; he doesn't have a lineage that goes back to the barbecue dynasties of North Carolina or Kansas City. Nor does he compensate with a lot of made-up lore. He's a large man, given to vests, who was working in high-end security when he got a pill-shaped Weber smoker as a present from his mother. Soon he'd grown fascinated with cooking big pieces of meat as slowly as possible. Preparing fourteen- to sixteen-hour brisket for family and friends had eventually led to him showing up at divey bars with large platters wrapped in butcher paper and to being known as the guy who brought the fall-apart brisket at just the right time in the night.

That acclaim had led him to open Hometown Bar-B-Que in the Red Hook section of Brooklyn in 2013. An empty slab of a warehouse when he bought it, Durney transformed the depot into a warm venue encased in wood from an old barn with a pair of tumbler-stained bars and long communal tables where customers eat. The focal point of the restaurant is the slightly elevated counter where the meats are cut, whacked, chopped, or pulled and served up with a choice of sides like corn bread and mac and cheese on half-sheet trays lined with paper. Now, at noon, he watches a line of early arrivers snake toward the counter where the brisket, crunchy with layers of delicious bark, was being sliced. Plenty waited for their appetites to kick in and drank beer. A child's birthday party was taking place in the room that usually acted as a second bar, and Durney went to check in on the preparations. As he sat at the bar, Conlon looked as content as any of them.

When I'd met Conlon a couple of hours earlier, he was just finishing his all-night vigil. It was a bit like seeing a miner after he gets off his shift. His work radiated from him. Conlon wore thick insulated overalls, the bib coming up over a thick fleece hoodie, and a tan, heavy cotton welder's jacket. His hands were covered in soot, and the tip of his salt-and-pepper beard had been turned a smoke-tinged yellow by his nightly task. It made sense to be so protected. Red Hook juts into New York Bay—the Statue of Liberty is almost a resident—and in February, a wind, the very breath of the Atlantic in winter, blows through the Verrazano Narrows, meeting its first resistance in the row houses outside along Van Brunt Street.

As we sipped shots, Conlon told me a bit of his story. His fascination with firewood cooking had started when he worked at an Italian restaurant where everything, the pizzas and wood-roasted vegetables, came out of the embers. What he did now, tending each of the fireboxes on three drum smokers mounted on trailer wheels while rarely lifting the hood to avoid losing the trapped heat, was like a virtuoso performance of the art of firewood cooking. Conlon put it in terms of the stoves he once toiled over. "Our stove dials are those dampers," he said. Open too little and the smoke doesn't move and coax flavor, melt fat, and tenderize muscle to fall-apart softness. Open too wide and those dampers spur the flames and ruin the meat. "It's a very pure way of cooking," he explained.

There was a delight in his occupation and a seriousness of purpose that were definitely at odds with what I'd expected from Brooklyn. No realm of artisanal food is more mythologized than this expanse across the East River from Manhattan. Comedians skewer the heightened level of self-preoccupation; the bushy beard many of the local men have made popular has become an easy gag. Rather than a place that sits between Queens and Jamaica Bay, it has become the launching point for the modern foodie stereotype as a skinny-jean-clad, slightly censorious individual telling the nation that those things we enjoy are somehow not stark or true enough, and that while they won't deny us the pleasure—if that must be the name for the sensation our commercially produced chocolate gives us—it's not what's considered on trend at Bedford Avenue and North Fourth Street.

If I were being honest, I'd admit I've come to verify, to get some scenes, jot down some notes that prove it's true. I've been here only two days, but already I see how different the reality is. At every opportunity, Brooklyn has turned the tables and questioned me. Conlon is not some dilettante, but a man completely committed to his task. On my first night I'd gone to get a drink in a small spot on DeKalb Avenue and immediately spotted rows of empty brown glass beer growlers, the standard receptacle of microbrewery fill-ups, ringing the room. I'd barely finished writing *Growler not for beer but for look* when it struck me like the observation of a prim prosecutor. It wasn't anyone in an Auden-in-Berlin haircut that was tut-tutting; it was me.

The elevated sense of work that artisanal Brooklyn is known for is something I'm relatively familiar with. I've had an espresso crafted for me in a Portland, Oregon, cubbyhole that could hold five people perhaps, watching as the sole employee leaned forcefully down with the tamper to create the firmest puck for the water to seep through. Across the country in Portland, Maine, I'd scattered handcrafted oyster crackers over a chowder, curious to taste what it meant to the chef to make the crackers rather than tuck a cellophane bag of them between bowl and plate. Both actions had the aim to create an added degree of flavor. But was it studied, calculated to impress, or was it real and authentic?

Being one subway stop from Manhattan elevates those concerns. The proximity magnifies everything. It can reduce any act to a stereotype; it can transform the tiniest halting endeavor in a Bushwick walk-up into an illustration of the newest trend. Seen from Williamsburg, where the cobbled streets with fire escapes open onto a vista of its neighbor, Manhattan seems to be peering over with a guarded curiosity. Perhaps even a nostalgia. A block-by-block map of Manhattan published in 1900 reveals the vitality of a city where brewers, malters, and pickle factories were woven in amid the fraternal orders, the public schools, the coal yards, the churches, the synagogues, the cigar manufacturers, and the ice merchants. Today, it is rare to sense any kind of food manufacture going on in Manhattan. Sure, walking down Mott Street, you're enveloped by the smell of freshly steamed pork buns, but, for the most part, artisanal foods have been displaced by the value of real estate. They're something to be displayed, not actually crafted, saluted in the form of a perfect tower of *macarons* from a Parisian pâtissier displayed in the food hall of the Plaza.

Over here in Brooklyn, I'm constantly being surprised by how food manufacture coexists with a real place. The Polonia Democratic Club & Ladies' Auxiliary on Grand Street seems impervious to the bakery displaying emmer bread nearby. A store like Stinky Bklyn, located amid the grocers and old-time butchers of Smith Street, can offer a selection of organic bitters, be nicely contrarian—UNATTENDED CHILDREN WILL BE GIVEN AN ESPRESSO AND A FREE PUPPY reads the sign—and still be a place where an employee is happy to slice the tiniest orders of hams from Allan Benton in Tennessee or Ronny and Beth Drennan in Kuttawa, Kentucky, for customers who want to have a tasting session at home. That's the kind of food-related enthusiasm that places Brooklyn within a new version of the American city. In Minneapolis, I've joined shoppers at the Mill City Farmer's Market near the remains of the great Washburn and Pillsbury mills shopping for vegetables (and the Salty Tart Bakery's cult favorite squash focaccia) while being serenaded by a German oompahpah band. In San Antonio, I've torn apart wood-fired pretzels in the crowd at Southerleigh Fine Food & Brewery, located in the repurposed shell of the city's historic Pearl Brewery. In the Walker's Point section of Milwaukee, amid the disused foundries and warehouses, I've joined the group sipping for nuances in the absinthe crafted at Great Lakes Distillery. The cobblestones from a different era often announce some delicious product today. Sort of like Van Brunt Street.

Durney makes what he calls "New York–style" barbecue. While there are certain foods that the name of the city serves to modify—bagels, pizza, or an egg cream with the ribbon of chocolate syrup blended into the seltzer and milk—barbecue is not among them. There are people in cities where barbecue has lore and tradition, where the designation might be met with a wry nod. North Carolina barbecue depends on the contrast between the tang of vinegar and the richness of hand-chopped pork. In Memphis there might be the hint of molasses in the sauce, while the more tomato-based sauce Texans favor contributes a lush layer to the brick-sized ribs. In Kansas City the tradition is linked to the storied stockyards, and they'd likely joke about the resonance of New York–style 'cue.

It is, however, an honest term. Both Durney and Conlon are from Brooklyn, and as we stand at the bar, waiting for food to come out—Durney wants me to taste everything—they try to give me a sense of where their barbecue is rooted. Brooklyn is a big place. "You'd have to be looking for Old Mill Basin to find it," says Durney of the neighborhood where he was raised. "I'm from Canarsie," says Conlon. "Nearby, though we never ran into each other." There had been a sense of that pride in their home borough earlier as Durney, Conlon, and I had stood in the outdoor yard they keep the smokers in and Durney saluted great and inspiring barbecue joints such as the Skylight Inn BBQ in Ayden, North Carolina, and Louie Mueller Barbecue in Taylor, Texas. Each one perfected a style that was regional, and if they didn't reach all the way back to the time of cattle drives, they certainly captured something particular in the way they utilized acidity and sweetness, crunch and crackling, to parry the profound satisfactions of slowly cooked meat.

Hometown's barbecue had been shaped by a different set of local conditions. Durney had barely settled on his restaurant location when Hurricane Sandy made landfall. The low-lying streets of Red Hook were flooded and power cut off. For the first few weeks after the storm, Durney turned his attention to feeding sanitation workers, cleanup crews, church members, and locals who couldn't use their own kitchens. In a way, those days in October 2012 forced him to take that last step from amateur to professional. "It was the first time I'd cooked fifty racks of ribs," he says. "I had to."

One of the T-shirted employees comes out with a half-sheet tray lined with paper and a few different cuts of what Hometown serves. The moist brisket represented the glory of barbecue—the all-night process had transformed the cut into a carnivore's delight of luxuriant meat encased in caramelized bark. This was what time did, breaking down muscle and fat into profoundly satisfying mouthfuls, its only seasoning a hint of the backcountry. The other meats I tried tossed a little creativity into the mix: pineapple juice, hot sauce, and cider vinegar teased out the funky note in the pulled pork; a good rubbing of Jamaican jerk spices and cherrywood smoke lingered on the ribs (thankfully I got only two bones of the rack). Durney was particularly proud of his lamb belly. "It should be like a stacked burger," he says, pointing out meat and fat in equal measure. He served it as a banh mi, slung in a toasted half-baguette, its salt, pepper, and turbinado sugar seasoning, plus oak wood smoke tamed by the ribbons of pickled daikon and a handful of cilantro. "It's not a classic banh mi," Durney says as he watches me take a mouthful. "If I put pâté in there, I'd kill people."

The last thing I eat is the pork belly pastrami, a wedge of deeply marbled Heritage Farms Cheshire pork next to a spear of sour pickle that's been finished in a brine made with upstate rye whiskey. The snap of its texture and the bracing tang of its flavor make the trembling meat seem even more lush. "That's a great combination," I say. "And a great pickle."

"You should go meet Shamus," Durney says. "He's a real Brooklyn artisan."

"Where did you meet him?" I ask, interested how the connections between businesses in that particular circle are made.

Durney makes a face, thinking. "Probably in a bar," he says as he starts scrolling through his cell phone to give me his contact info.

PERHAPS BECAUSE I ALSO LIVE in a city so often reduced to parody—Angelenos are obsessed with cleanses and getting to the gym by dawn and only care about a restaurant if it caters to stars—I sense myself gradually rejecting the satirical image of Brooklyn as a place of the overcurated life. Still, I have certain preconceptions of a Brooklyn pickle maker, and I have to remind myself of that the following day when I meet up with Shamus Jones, the founder-owner of Brooklyn Brine at his pickling plant off Fourth Avenue in Gowanus. Jones does not look like some minor character in a Joseph Mitchell story. He is rail-thin; despite the cold, he is wearing a kelly-green T-shirt that reads ITHACA IS GORGES under a skimpy windbreaker. He is talking on the phone about a possible new location even as he stands in his old one, a brick building—with a worn wood barrel outside—he has outgrown. He has been making pickles for less than a decade and has already moved his manufacturing plant three times.

Jones is a new kind of traditionalist. He plays guitar in the punk band Psychic Limb specializing in thirty-second grind—a genre so demanding a song can't often go past a half minute because the players are physically spent. His pickles are preserved by the acidity of vinegar often embellished with an added natural ingredient such as maple syrup for flavor. Still, as he leads me into a back room where about ten people are working, I see all the signs of traditional manufacture. There are cases of raw, fresh washed cucumbers, and against the far wall a mass of open jars into which workers measure the prescribed amount of mustard seeds, caraway seeds, dill flowers, and peeled garlic. Over beers around the corner at Pickle Shack, the restaurant he recently opened, Jones tells me how he got into the business. Raised in Brooklyn by a single mom, he had moved out to Seattle with her as a teenager and was soon working in restaurants and learning the art of preservation. He started there at Cafe Flora, a neighborhood gem located in a former Laundromat, and then at Carmelita, a vegetarian restaurant that was an early proponent of foraging known for its nettle and Manchego soup.

Jones recalls being struck by the philosophy of a restaurant being a part of the community. Cafe Flora had a well-established food donation program with the neighboring Bailey-Boushay House—the nation's first AIDS hospice—and a serious composting program to cut down on waste. "On my first day I was handed a five-gallon bucket to keep at my station. I didn't know what it was for until they explained it was for all the peelings."

Both restaurants, too, saw their creative pickling programs as grounded in a no-waste approach. At its heart, pickling is a means of extending the life of something raw and delicious, preventing the onset of that thin film of sliminess that marks the start of putrefaction. As a young cook, Jones was struck by the worldwide breadth of the techniques he was learning. Garnishing an appetizer with pickled watermelon rind delved into the American pantry; those salty preserved carrots and lemons that might get tossed into a stew came from a North African one. He wasn't sure what tradition pickling garlic scapes, the loose tangle of stalks that rise from fresh bulbs, came from, but they became a surprisingly delicate counterpoint to the chalky purity of a chèvre.

That was the preparation that began his artisanal career when he returned to Brooklyn in 2007. In typical Brooklyn fashion, this involved a crisis (loss of a job), a walk-up (where he hatched a plan), and a borrowed restaurant kitchen where he briefly pickled at night. It also resulted in the sort of crazy acceleration that a food artisan can experience in Brooklyn. The business started small in 2009. The first sale was to Urban Rustic, a Williamsburg food emporium; the first returned order came from Murray's, the Manhattan cheese store. Jones's Middle Eastern pickle fascination was in full flower and he was making *rass el hanout* pickled eggplant with preserved lemons. "Murray's called," Jones tells me as we sit at the counter at Pickle Shack. With a self-aware smile he recalls them saying, " 'Can you take these back?' It was so precious." He laughs, not surprised now that they didn't sell. Still, only two years after starting, Williams-Sonoma called, asking him if he could deliver twenty-two thousand jars. "We didn't think we could do it," Jones says, sounding slightly amazed at what that had entailed, "but we did."

It's easy to understand the appeal, both for a large company and for customers prepared to spend a couple of bucks more on a jar than they usually would. A Brooklyn pickle has all the sharpness of a real American experience compressed in a single ingredient. Usually food-related Americana has rustic roots, but the pickle is a city art. In *A Walker in the City,* Alfred Kazin's book-length memoir of his 1930s Brooklyn childhood, pickles are Proustian objects, present (all too present when they symbolize the constricted world he seeks to leave) at the Rockaway Avenue subway stop, his family's Sabbath table, and, forcefully, along the Blake Avenue street market, where "torches over the pushcarts hold in a single breath of yellow flame the acid smell of half-sour pickles and herring floating in their briny barrels."

This was a world of loudly sung Yiddish folk songs and Socialist hymns where raw cucumbers were hauled from Long Island farms and all that was required to preserve them was a barrel, water, and salt. Plenty of the last one. The salty solution inhibited any nasty bacteria from gaining a foothold and allowed naturally occurring lactobacillus (incredibly hardy and resistant to salt) to ferment the pickles' natural sugars and create the acidic environment that preserved them in the classic cloudy brine. Whether in the tenements of the Lower East Side or the tightly packed streets of Brownsville, for Jewish immigrants the pickle was the original value-added product.

I am reminded of those beginnings the next day when, after a rough start, I visit Ba-Tampte, a classic—perhaps even timeless—Brooklyn pickle manufacturer. I'm up early, dressed, and ready in the lobby of the Aloft Hotel near Borough Hall waiting for a livery car. Fifteen minutes after the pickup time I'm still waiting when my phone starts to buzz with texts, then rings. My driver calls me. "Traffic," he shouts over a squawking dispatcher, "another car's getting yah." Minutes later I'm bouncing in the tufted back-seat of a lime-scented Reagan-era town car as we sweep around the Soldiers' and Sailors' Memorial Arch that marks the entrance to Prospect Park. We head south along streets that grow progressively narrower until canyons of houses lined with weatherproof siding give way to body shops, salvage yards, and, eventually, the Brooklyn Terminal Market, a low-slung building opened by Mayor La Guardia in 1937 in outer Canarsie.

A few trucks idle in the central yard while I enter the doorway marked Ba-Tampte where a group of four men are cleaning a long series of conveyor belts with pressured steam rods. I ask for Mr. Howie Silberstein, and one of the men leaves to find him. A few minutes later I hear a voice coming from somewhere amid the massive urns of stored pickles. "I'll be right there." Eventually he appears. A tall man, almost seventy years of age, wearing jeans and a flannel shirt, who has some bad news for me. "We're not making pickles at the moment," he says. "We're cleaning the bottling line before we start the Passover pickles."

"They're different?"

There's a moment's hesitation as if he's deciding whether the details are worth getting into, and then he does. The men, meanwhile, are looking for mustard seeds, using the steam guns and pressurized hoses to flush out the vats and the mechanized belt the jars run on. Silberstein explains that the little seeds are a key flavoring component of the usual brine, but he doesn't—he can't—use them for the pickles that will be on the Passover table. "It's not that mustard seed is *chametz,*" he says, giving the Hebrew word for the leavened foods that are prohibited in the house during Passover, "but there's some question about whether it's not."

I like that. The little orb of flavoring spice playing a role in the heritage of rabbinic responsa. In a way it becomes seasonal pickling, except the season is halachically decreed. Ba-Tampte's kosher certification is not in doubt. It's closed on the Sabbath, and there's a full-time rabbi on premises. But why run the risk? We've barely begun to speak when, spotting something, Silberstein suddenly bends down and presses his finger to the cement floor. "This," he says, showing me his index finger to which a tiny mustard seed has become affixed, "is what we're looking for."

He motions me into the meeting area, a faded paneled room with files in different stacks and a feeding bowl and litter box for the company cat. There's plenty to explain, and he takes evident pleasure in doing so. *Batampte* means "tasty" in Yiddish. The company uses only Kirby cucumbers. The emphasis on seasoning spices came around centuries ago when they figured out a fermented pickle all by itself was kind of tasteless. With that for background he settles into the Silberstein part of the story. Around us are pictures of pickle stands his father and grandfather kept, which he intermittently takes down to give me a closer look. "We were at the Essex Street Market, too," he says. "It opened in 1940 because Mayor La Guardia's idea was to get the pushcarts off the streets." Over the next half hour, and with the cat sleeping nearby, I get to hear the full Silberstein tale. Establishing the company in 1953 was almost the crowning point of a saga. Howie's great-grandfather Aaron had brought the pickling art over from Chernivtsi, a town in the Romanian backwater of the Austro-Hungarian Empire, and passed it on to his son Reuven, who passed it on to his sons Meyer and Abraham (known as "Boommie"), who kept a pickle stand near Nathan's Famous hot dog stand by the Scooter Palace bumper car ride on Coney Island, who passed it on to Barry and his younger brother Howie, who grew up standing by the barrels making retail sales. "Eight cents each or three for a quarter," he says, his face broadening into a smile. It takes me a moment to do the multiplication and then

I smile back, realizing the mischief. "Hey," he says, "no one asked for the penny back."

Outside on the closed-down bottling line, he lets me taste from a large tub of brine. I stick a finger in. This isn't some little timid brine, but rather one that sucks the moisture from my tongue and cheeks the moment I taste it. There's no way a pathogen could operate here. That salinity is enough to preserve the sweet Kirby cucumbers he uses, but, because he doesn't stop the process with heat treatment, it keeps right on going. Chilling is the only way to keep these pickles stable. "This is traditional," he says, emphasizing every vowel and growing animated as he recalls people who don't follow instructions to keep the jars refrigerated in their stores. "I've pulled accounts from guys who keep cases in their van on a hot day and then call complaining the jars are fizzing. Of course they are," he says, punching the air for emphasis. "A jar is just a miniature barrel!"

I could bask in Silberstein's company. There's nothing hesitant about him. All those accounts to keep track of to make sure they're taking proper care of his cukes must be the same as his father and grandfather knew. But there's a modern version of that pride, too. I'd sensed it back at the Pickle Shack when Shamus Jones was telling me about the complicated manufacture of the pickles he makes flavored with an India pale ale beer from Dogfish Head Brewery. "You don't just pour in the beer," he'd said. "They have to mount the hop oil on a tincture and we work from that."

This was a long way from Ba-Tampte's ageless flavoring combination of garlic and dill. But ultimately, both were recipes. The greater difference was that Silberstein was continuing a tradition and Jones had started without one. After warm farewells, I am back in the town car, and as the driver blends into traffic, the modern Brooklyn artisan's readiness to build something new strikes me as noble and worthwhile. By experimenting with flavors, Brooklyn Brine's Jones is being true to the experiences that brought him to pickling and enriching a tradition with creativity. Food is a generous realm, opening up and offering an identity to those who choose to find it there, and it had granted that to the Silbersteins just as it has for Jones. The driver hits a seam in traffic, and the streets of Canarsie are now blowing by. The key contrast is that the modern artisan starts the journey from a solitary place. There is no ancestral dimension, no Coney Island, no Chernivtsi, no elder's hand. Brooklyn artisans are on their own, finding and developing a skill that eventually, maybe, will grant a greater meaning to their lives.

I SPEND THE NEXT FEW days in a flurry of eating. I taste seasonal vermouth with a trace of butternut squash glowing somewhere amid the botanical notes and, at the Widow Jane Distillery, sip rye whiskey made with water from the limestone mine in upstate Rosendale. That distillery shares a redbrick building with Cacao Prieto, a chocolate maker specializing in bars made from single-origin Dominican beans. I join a tour with Brazilians, Germans, and curious New Yorkers and we all listen as the guide explains the process, putting particular emphasis on the mix of vintage and modern machinery the chocolate company uses. The ball roaster that brings the beans of intense dark doneness is a hundred years old, but the winnowing machine, a built-in-house apparatus of see-through tubing that takes the husks and leaves the nibs for grinding, looks like something from *The Jetsons*. After the tour, I'm standing in the inner yard between the buildings eating a few squares of the fabulous, sea-salt-dusted chocolate when a bunch of heritage chickens appear and start pecking around me. "Well, I guess I'm in Brooklyn," I say softly to myself as if the scene were now complete.

It isn't that the borough didn't have an artisanal food tradition before becoming the focal point of the modern movement. In 1865, when Olmstead and Vaux laid out the plans for Prospect Park, they included a working dairy, intended to complete the bucolic suggestion of deer in the paddock and sheep on the Long Meadow. Even amid the urban dystopia of the late 1970s, the Greenmarket on Fulton Street was one among a handful of markets in the entire city to unite traditional truck garden suppliers and activists growing vegetables in empty lots in one setting. Damascus Bakery on Atlantic Avenue has been selling still-warm lengths of oven-singed lavash flatbread since the Halaby family opened the business in 1930.

To join that line is almost to participate in a realm of Brooklyn that has no time for or interest in (perhaps not even any knowledge of) the artisanal fame of the borough. Despite the red-hot intensity of the pursuit of new old flavors, it is quite easy to go about one's day oblivious that the realm even exists. The pretzel seller on Havemeyer Street is breathing steam through a scarf and moving his feet to a jerky CD of the Fania All-Stars. On Saturday evenings, vespers are sung under the onion domes of the Eastern Orthodox Cathedral of the Transfiguration, the Muslim call to prayer sounds from loudspeakers above the bookstores at crowded intersections, and a no-nonsense staccato that might serve as the borough's signature is set by the rapid paces of the black-hatted men leaving tiny shuls along Eighth Street after *Maariv* services. Later at night, the multi-culti crowd ensconced in booths at Junior's at Flatbush and DeKalb

Avenues are content to share big slices of good cheesecake amid a time-burnished display of seats from the Dodgers' old Ebbets Field home and headshots of locals who made it big. Dimple-cheeked actor Tony Danza is there on the celebrity wall and so is rapper Notorious B.I.G.

But artisanal Brooklyn is ever-present, too, as if seeking to keep the artisanal conversation serious. The big questions that hover around small-batch food manufacture become very clear here in all the cacophony a great city provides. Does the "intention" with which something is made really represent a modern value? If the IPAs start crowding out the Goya canned beans on the bodega shelves, does that mean the neighborhood is ruined? Why must effort be painted as something dilettantish? How do you deal with the tired chestnuts about hipsters and remain focused on developing your craft?

I realize how easy it is to slip into the tone of ribbing one morning near the end of my trip when I'm walking through the streets of Williamsburg on my way to a meeting with Garrett Oliver at Brooklyn Brewery on Eleventh Street. I'm early and ambling northward, and I've just rounded the corner from Berry Street when I catch sight of the full Williamsburg beard. Perhaps it's how the guy is backlit by the river light, but I feel like an ornithologist spotting a rare breed in its natural setting. It is magnificent, lush, impervious to opinion, shattering notions of grooming even as the curled tips of the mustache indicate how painstakingly it is maintained. Minutes later I am inside Mast Brothers, a famed chocolate maker where stacked bags of beans are incorporated into a decor that's as monochrome in its look and as intentional in its display as an Apple store. I'm having a perfectly nice time taking in the scene when I turn around and, beyond the bags and the counter, I see the faces of two white-capped young workers peering with an intensity I recognize into a drum where granite millstones turn. I've seen that face before—drawn, intent, scared of fucking up. I've *had* that face when I first worked in kitchens. It's the face when you realize you're in a place where they're moving at a speed you don't have, capable of communicating rapidly using terms you don't know and of creating things you still can't. That, too, is an artisanal moment because it's often the starting point of the discipline. It's not flee or fight; it's flee or learn.

Going from joking thoughts about beards to more fundamental ones about the nature of work feels a bit like whiplash. In a way I'm glad to be going to Brooklyn Brewery, one of the earliest businesses to reestablish the identity of the borough. Instead of commentary it seems like a primary source. Founders Steve Hindy and Tom Potter launched the brand with its sweeping Milton Glaser–designed logo in 1988. Then they had the help of Bill Moeller, a senior brewer at Philadelphia's C. Schmidt's Brewery who liked their idea of reinvigorating what had once been Brooklyn's vital beer scene. But the beer was being brewed in Utica. In 1995, they leased the former Hecla Ironworks on North Eleventh Street, a few blocks from the East River, in what was then a very run-down Williamsburg, having hired Garrett Oliver to be the brewmaster. At the appointed time, I ring the bell on the side of the large building and am buzzed in.

A tall, elegant man wearing wide-wale corduroy pants and a loden-colored quilted vest, Oliver strikes me as if he'd fit in perfectly at a traditional European hunting lodge. We meet in the brewery's offices, an open-floor plan with sound-baffling partitions, and he leads me down the metal staircase for a rapid tour of the operation. There are massive kettles and drums to be seen as we step over hoses discharging yeasty foam into grates. At the tasting room, a large space with long tables, he pulls two short pours of *saison* for us—"It's good for early drinking," he says with a laugh—and pauses in front of a display of antique glass beer bottles. "Brooklyn was one of the centers of American brewing," he says, taking a sip of his beer and leading me back upstairs so we can talk.

That tradition was due to the size of the German population and to the quality of the water—something Manhattan hadn't been able to boast about until 1842 and the building of the Croton Aqueduct. By then the city had laid down the seeds of a heritage that blossomed into beloved local brands such as Schaefer, Excelsior, and Rheingold. It would have been hard to predict in 1960, when Schaefer was still producing eighteen million barrels (one-fifth of the nation's lager), that its days were numbered. By 1975, it was down to five million barrels, and the next year it closed. It was followed by Rheingold. The brewery that had once sponsored the hugely popular Miss Rheingold competition staggered through a brief Pepsi ownership and the ignominy of being sold to Chock Full o' Nuts for one dollar, and in 1976, after 121 years in Brooklyn, all but one of them at the same Bushwick location, finally turned off the brew kettles and closed its gates.

Garrett came out of the rebirth of American brewing. Peppering his talk with the vivid details of a natural raconteur, he leads me through the steps of his learning as we sit at a conference table. He'd discovered the rich complexity of English ales while working in London in the early '80s, and after his return to New York, he set about trying to re-create them by learning to home brew.

"The wonderful geekery of modern brewing didn't yet exist," he said. "I'd go to Milan Lab down on Spring Street, which was a holdover from Prohibition winemaking, and I'd be lucky to find some single-strain yeast not infected with who knows what." He helped found the New York City Homebrewers Guild, which gave him an opportunity to improve with other amateurs. In 1992, he became brewmaster at the Manhattan Brewing Company, a short-lived multitap endeavor in a repurposed Con Ed substation at Thompson and Broome in SoHo where he worked for almost five years, half of them under original co-owner Richard Wrigley, an Englishman with deep roots in British brewing.

Two years later, Garrett attended a London conference of the world's great ale brewers, where he'd been invited to be among those presenting their beers. He still remembers the nerves he felt on the day he left as he sat on the runway, a miniature keg of the hoppy ale he'd started to brew at the Manhattan Brewery Company packed overhead. The excitement is palpable in his voice all these years later. "Fuller's, Young's, Bass, these guys were all my heroes," he says, recalling the moment with a laugh, "and here I am, a black kid from Queens taking them my beer to taste."

Though Garrett is recounting an anecdote and not making a broader point, the mention of his ethnicity cuts through—with a jolting speed I've come to almost expect from Brooklyn—to a fundamental aspect of the artisanal experience. The artisanal food world can seem very white at times. That observation is there for anyone to make. Certainly to consider this new realm of flavor through the prism of the African American experience is to have it shift on its axis. A desire to return to old values can ring hollow. Facts can suddenly pertain to a very different narrative. To slaves, the Ohio River was not an opportunity to settle land or send barrels of whiskey to market but the way they were transported to the plantations of the Deep South.

Many slaves learned crafts. For some it was those of the plantation home such as weaving; others, responsible for the grounds, were skilled carpenters, while still others worked at industrial activities in forges or, like young Frederick Douglass, became shipyard caulkers. But skill was all they had. Because their labor had no monetary value, it was impossible to equate an acquired craft with any measure of autonomy. As Eugene Genovese put it in *Roll, Jordan Roll,* "The term slave-artisan is a contradiction in terms." Adding a very real poignancy was the artisanal pride felt by African American coopers, blacksmiths, and others despite being unable to claim the freedom that came with a trade. Genovese cites a case of plantation carpenters who, ordered to construct a quick seasonal outbuilding, built a totally different structure but did it so well the furious planter couldn't bring himself to tear it down. In a twist, a measure of revenge was exacted by the excellence of the work.

In that light, merely being able to put one's hand on a favored piece of equipment becomes a moment that cannot be taken for granted, just as history can serve as a reminder that being an artisan is a privilege some Americans never had. The artisanal ferment in Brooklyn becomes thrilling when interpreted within that context. It is the opportunity seized, the responsibility assumed, the independence honored. In fact, this world of craft and recaptured flavors has succeeded to an extent it has spun off new kinds of hurdles. One of them is getting too caught up in the story of Brooklyn, a chronicle that demands there always be a new young person with a new dream lugging the raw material toward the walk-up in the opening scene. What happens when the idea succeeds? What happens when you're no longer the visionary in the home kitchen? What happens when you emphasize your newness by disparaging what came before? If I had any doubt about how that rejection could sting, it's made clear when I ask Garrett how he feels about tiny new breweries that today define themselves against the scale of Brooklyn Brewery. "It bugs the hell out of me," he answers, growing very serious. "Not being able to grow, it's like telling a kid you can only love them when they're five and cute."

The sentence knocks around my head for the rest of the day. I'm leaving tomorrow, and as evening comes on, I soak up the images of people meeting in the small coffee shops, moving briskly through the streets as they shop for dinner. I hit a few of the foodie markets I've made a note of, but there's nothing pretentious at any of them. At the Greene Grape on Fulton Street the butcher is having a conversation with a shopper on where to find tongue. "Not so easy to find," he says with a tone one of Kazin's pushcart merchants might have saved for the shopper rummaging through a barrel for the ideal half-sour pickle. This is the everyday, not the parody. At Brooklyn Fare on Schermerhorn Street, the checkout line is long with people carrying wrapped wedges of cheese and packs of coiled, plump sausages. I buy some flour-dusted hearth-baked rolls and look at my watch. If I rush, I can still make it up Smith Street to Stinky Bklyn to buy a few slices of that ham from the Drennans for the plane.

TO SCALE

T*artine Bread,* the book by Chad Robertson (co-owner of Tartine Bakery in San Francisco), was published in 2010 and is an ode to naturally leavened bread. Perhaps *ode* is the wrong word, and *manual* more apt. Like Jacques Pépin's mid-'70s *La Technique* did for a previous generation, Robertson's tome keeps to clear instruction and step-by-step imagery to teach readers how to work like pros. Pictures of Pépin's hands teach the difference between *macédoine* and *brunoise;* Robertson's tatted forearms show just how to fashion magnificent loaves. "Learning a craft is as much about copying as it is about understanding, as much visual as it is intellectual," Robertson writes in the introduction, striking the tuning fork on a work that throughout gamely attempts to put what a skilled baker feels with the tip of his fingers into written form. "Using the bench knife and one hand, work each piece of dough into a round shape," he instructs about shaping the bread. "Tension builds when the dough slightly anchors to the work surface while you rotate it. By the end of the shaping, the dough should have a taut, smooth outer surface."

One of the great pleasures of visiting the San Francisco bakery itself, which Robertson and his wife, pastry chef Elisabeth Prueitt, opened in 2002, is how this knowledge, touch, métier suffuses the atmosphere. The business occupies a busy corner at Eighteenth and Guerrero in the city's Mission District. If you peer in the windows on Eighteenth Street, you can see skilled bakers handling the impossibly wet dough Robertson likes to work with, taking one of the proofing mounds of dough from the workbench, performing those fabulous fast folds that look like diapering a particularly pudgy baby before they gracefully invert the blob into a proofing basket heavily dusted with a mix of wheat and rice flour where it will continue to rise.

For any of the many people whom the book has inspired to bake, this is like seeing the perfect form they are after. Of course, they don't have Tartine's commercial oven, but Robertson's suggestion of using a highly heated Dutch oven to replicate the ideal steaming and searing heat is attainable for the home baker. I've been on the receiving end of the results when my friend Michael Mullen, a Tartinist through and through, drives over to share a loaf that worked out particularly well. Most of those taking their place in line at four in the afternoon outside Tartine are admitting they were never going to master those moves. It's just as meaningful to be here, inching toward the bakery's door. You finally step over the entranceway, that incongruous inlay of the name "Carl's," vestige of a former business, and creep up to the pastry display that's the last sweet agony before you can order the bread. It's going to be hard to not ask for those cacao-nib-filled rochers that rise like soft-serve ice cream cast in the form of meringue. The big hotel pan of fruit-covered brioche bread pudding, they had to go and put it right there? You can smell the bread now.

You're so close. One person takes the order and another brings the bread in a brown paper bag with handles. The woman in front of you—she looks like she knows what she's talking about. She's ordering the sprouted wheat loaf. Maybe that's the one.

The feeling of getting a loaf handed over to you is so exciting you'll easily forget that you've been in line for well over an hour. The reason for the wait is that Robertson has always baked in what he calls "real time." Unlike a more traditional bakery that opens in the morning with a full display of loaves and pastries and then sells them throughout the day, Tartine bakes what suits the moment, croissants in the morning, gorgeous dark thick-crusted boules in the afternoon. A person could quibble that a customer's time has become subservient to the baker's time in this arrangement and that being allowed to buy that night's bread, say, during lunch break, would not represent some major concession. The logical reply would be that no one is being forced to buy this bread. There are many bakeries, not to mention pretty decent parbaked loaves you can toss in the oven straight from your freezer any time you please.

Of course there is far more than a transaction over a staple taking place. The customers are investing their time and anticipatory energy into getting something unique. The bakers are being asked to represent the highest standards of craft in Tartine's bread, with its dark, swelling, jagged crust, and coming through. Together, between the time of one and the skill of the other, a field of excellence is being created that suffuses the creeping queue, the glass-fronted display case, the quirky interior with its zinc counter and scattering of tables. Money changes hands, but that doesn't mean there isn't another, more abstract and wonderful exchange.

All our go-to spots contain this kind of energy economy. The barrista who knows your name and your drink is giving you far more than coffee. But Tartine was limited to about 250 loaves a day—predicated on quality, it was prepared to lose customers rather than serve bread differently from the way they wanted. Then the business went in a different direction. The story that Tartine was becoming a part of Blue Bottle Coffee appeared in the *New York Times* on April 20, 2015. In and of itself, it wasn't momentous news. Propelled by $26 million in venture capital money, Blue Bottle was undergoing an expansion that would mean more locations and, now, more pastries. The experience the roaster offered, one of almost ceremonial simplicity occurring around a display of digital scales and Chemex filters, was not likely to change. It's not like this was corrupting Robertson, either. He was already thinking of opening a bakery in Tokyo.

The comment thread on the story is interesting for the glimpse it gives into what happens when an artisan changes how he or she operates. This being San Francisco, there are some instructive opinions on the importance of water in baking, but the majority of opinions are unremittingly negative. One person quotes the Once-ler from *The Lorax*—"I had to grow bigger. So bigger I got"—others toss around the accusation that the new partners will now "go corporate." Admittedly, there is some instructive back-and-forth on the wisdom of cutting a Poilâne boule while still warm, though the overall tone is *spurned*. "This news saddens me," one reader writes, suggesting that waiting in line at Eighteenth and Guerrero was never meant to lead to this. Robertson had probably been hearing many comments like that. When I reached him on the phone several months after the news, he was enthusiastic about what the future held and looking forward to giving his team of bakers new challenges. But he was tired of parsing what expansion meant. "I'm so beyond having people react emotionally to what we're doing," he said, sounding like a man who couldn't bear to hear one more anguished complaint.

Undoubtedly he was tired of the subject, and perhaps I came across as too inquisitive about the particular juncture he was at—craft intersecting with rapid growth. In any case, as texts suggesting meeting times never came and an e-mail with a formal interview request went unanswered, I started to develop my own opinions just from the kind of silence a journalist recognizes as lack of interest. Figures like $26 million ($70 million in a subsequent funding round) are very hard to fit into the artisanal narrative. Blue Bottle was acknowledging this when it blogged that despite the developments, it remained the same company that had started roasting in "a 186-square-foot potting shed." The artisanal story may be predicated on being small and idealistic, but another narrative was taking place here, this one hinged on being big and realistic. In a way, Robertson's silence went to the heart of the matter and spoke to an unease I'd been feeling for a while. Even as it rose and grew, the artisanal movement was caught—like some airship tied down by relentless foes—in a series of false contrasts. Growth meant "going corporate." If you were big, then you were no longer indie and, once mainstream, you couldn't possibly be good. If I had to whack my way out of that thicket, I probably wouldn't want to talk, either.

EVENTUALLY, THE SITUATION RESOLVED ITSELF when the merger was called off eight months later in December 2015. "Our individual plans would be better served pursuing them independently," Robertson said in a statement. I could see how that would be the case. The Blue Bottle experience has a certain starkness—perhaps it's all the white space—that isolates the quiet technique that goes into crafting the pour and elevates the moment of drinking the coffee. No one here ever seems rushed. Embodying a great modern bakery, the Tartine experience takes place in close quarters: there's the happy wait, the kibitzing, and, most of all, the pure pleasure from warm bread.

It is phenomenal. Each loaf, from the fundamental country one to the sesame-crusted version to those made of ancient grains like durum, which the Tartine bakers have recently grown enthusiastic about, is baked darkly, erupting where it was scored with a knowing hand. The flour that was used to dust the proofing basket creates a delicious coating against the elevated ridges the baking process produces. When torn open—and most people do tear it—the crust looks like barbecue bark, the darker color of the crust growing gradually lighter as it descends toward an open crumb bearing the tang of natural leavening—but not pushed so far it overshadows the revelation that this is what skill can create, using only flour, water, sea salt, and fire.

Robertson devised the recipe in a six-year process of almost monastic discipline when in 1995 he and Elisabeth moved into a farmhouse in the coastal town of Point Reyes north of San Francisco. They'd met as students at the Culinary Institute of America in upstate New York where Prueitt, a former photojournalist, had detected a kindred independent spirit in her West Texan classmate. Together they'd gone to work in the Berkshires; Prueitt had continued developing her baking skills at Canyon Ranch in Lenox, Massachusetts, while Robertson had begun his exploration of naturally leavened bread at Richard Bourdon's shop, Berkshire Mountain Bakery, a few miles down Route 183 in Housatonic. Bourdon is a Canadian who along with Michael

London in Saratoga Springs, New York, and the legendary Helen and Jules Rabin (they'd been baking hand-formed loaves in their fieldstone oven in Vermont since 1977) compose a cell of natural fermenters in the Northeast. When Robertson's time with him was over, Bourdon had encouraged Robertson and Prueitt to continue learning in France. Their sojourn in Provence and Savoie was all about wood-fire baking, ancient grains, and the philosophy of *métier,* the mastery of a craft. Shortly after they returned, the couple moved into a clapboard house on a corner lot of Fourth Street in Point Reyes. They weren't quite renouncing civilization—Cowgirl Creamery was under construction on the same block—but they'd created a distance from it.

If one has to get away to focus on one's craft, Point Reyes is an ideal place. The seashore is a national park with a lighthouse on a rocky outcrop, and tule elk appear out of the fog that often hugs the coast. Along Drakes Bay, cliffs face a long strand of mouse-gray sand and crashing waves. With surfing as his sole means of relaxation, Chad had embarked on a severe form of self-education. He had renounced all technology, which meant he had no refrigeration and no mixer. For a while he even tried mixing the dough in a wood trough—the village baker of yore!—until he realized he was making something already hard even more difficult. In the focus on his work, in his severity and discipline, even in his awareness that the world must be kept at a distance for an internal journey to occur, this period in Robertson's life harkens back to the roots of the monastic experience.

Ever since the sixth century when Benedict of Nursia established the principles of monastic life, manual work and prayerful reading were seen as balancing each other, becoming what Benedictine nun Joan Chittister calls "opposite sides of the great coin of a life that is both holy and useful." The goods produced by the work were means to sustain the monastery, though, from amber Trappist ales to opaline Chartreuse liqueur made by the Carthusian order, they became early chiseled names in the rolls of artisanal foods. What's neat is seeing the process mirrored in more contemporary settings and how that wish to gain spiritual value from the work of one's hands can also be traced to the source of the modern American artisanal food movement.

I'd been reminded of that when sitting with Michael London in his Saratoga Springs patisserie, Mrs. London's. He's one of the pioneers of sourdough baking, respected for the quality of his loaves, well remembered at Zingerman's in Ann Arbor for providing five pounds of starter to begin Zingerman's Bakehouse and carrying it himself in an ice chest on a plane. In 1971, he was focused on baking as a form of revelation. He'd been coming at the matter head-on, as if creating the right setting for enlightenment was a necessary step. He was working at the Ananda East bakery on West Fourth Street, a narrow throughway between Sheridan Square and Sixth Avenue in Manhattan. He was studying the mystic works of Gurdjieff. He was baking wholesome if simple carrot brownies and soy apple cake. Nothing was happening. At some stage the New Yorker in him—he's the son of Brooklyn boxer Danny "Ripples" London—realized that in terms of craft, a well-made prune Danish displayed far more skill than any of the pastries he was making.

It was that awareness that led him to enroll in the apprenticeship program of the confectioners' union school in Long Island City, where he memorized William Sultan's *Practical Baking* and came in contact with German and Central European pastry masters. "Mr. Nackowitz taught me to pull strudel dough after class,"

London recalled to me in his still-strong New York accent, voice rising incredulously at how guarded the men could be about certain techniques. "These were guys who burnt their own sugar for the caramel in the pumpernickel and each one kept it hidden in a Maxwell House can in his locker."

As a union member he was sent forth into the Manhattan of the mid-1970s, a place where carriage trade French patisseries filled their windows with stacked towers of caramel-drizzled *choux* pastry puffs, and Hungarian bakeries broke for a goulash lunch about three in the morning. "Dumas, he made an incredible pecan ring," London recalled. "Greenberg's on Sixty-Eighth and Madison made the city's best cheese Danish. That's where I learned to make them."

Andrew Martin, the head baker at Greenberg's, had worked for twelve years at Mrs. Herbst's, a revered strudel palace on Third Avenue in Yorkville, and he brought those Old World standards to his work at Greenberg's. His *schnecken,* plump rolled sticky buns made from yeasted sour cream, were always fresh, nor did he skimp on the poppy seed filling in the *pressburger,* braided first cousin to the strudel. There was no sheeter to run the many types of dough through; Martin's barrel chest and strong arms came from taking a rolling pin to them all.

At first London, the lanky assistant, could barely keep up. "He told me, gently but firmly, 'Do you realize how slow you are?' 'Course, I did. Baking was becoming a meditative activity for me. I was starting to become self-conscious in the Gurdjieffian way where there's a double arrow. One arrow on the work and another arrow on yourself working. I did see how slow I was. I went home and I knew I could feel sorry for myself or I could do something about it." From that day on, he had Mr. Martin's apricot glaze ready for him before the baker even turned around to ask for it. It was a breakthrough for London, a form of enlightenment. He could be contemplative without working at a contemplative pace.

That understanding of baking as a craft where the outer execution has an inner value is a powerful theme in modern American baking. It's built on a paradox of the assumed discipline of craft being able to open up an inner spiritual space. Ed Espe Brown's 1970 classic, *The Tassajara Bread Book,* started in a cabin in the Big Sur Buddhist retreat, was written almost entirely in the *rōshi* tone Brown had internalized from years of Zen study: "Push down and forward," he writes about kneading, "centering the pushing through the heels of the hands." It's as if he's giving instructions for correct sitting posture during zazen meditation. Twenty years later in a 1991 book, *Brother Juniper's Bread Book: Slow Rise as Method and Metaphor,* then Brother Peter Reinhart (today an admired baking instructor and author) puts the crackle of a finished loaf in directly religious terms. "The sound of crust is like an icon," he writes, "not painted but baked."

In *Tartine Bread,* Robertson lyrically recalls the bread he was seeking to make during his Point Reyes years. "I wanted a deep auburn crust to shatter between the teeth, giving way to tender, pearlescent crumb. I wanted my baker's signature, the score made with a blade on top, to rise and fissure, and the crust to set with dangerous edges." The chapter on that period is bucolic in its tone, painting a picture of a young couple, living just steps from their wood-burning oven, establishing their craft and finding their place in the world. "Liz made pastry, using the oven early before it was fired hot or after it had cooled sufficiently," he writes. "We were soon bartering fresh bread for wild salmon, abalone, oysters, duck eggs, fresh fruit, and garden vegetables."

This is an idyllic world, small, perfect, and autonomous, so self-contained that it in fact presages the unease about growth when that phase eventually comes.

It is possible to stay small in the artisanal food world. When I think of what that looks like, I think of Brian Fredericksen who runs Ames Farm, a single-source raw honey enterprise, from an old dairy farm in the countryside west of Minneapolis. His entire business plan is to make tiny amounts of honey from very specific locations. "I use nine frame hives," he says. "That's hobby-level equipment." But it's where he places those hives that makes a difference. He keeps hives on the bluffs above the Minnesota River, in corners of untilled meadows, and at abandoned homesteads around the state. His goal is to capture that particular spot and, in his phrase, "the floral picture" that the bees with their honey are able to take.

These tiny amounts are often so nuanced they've made me question something very basic about flavor, namely the need to taste something intensely to think of it as authentic. Many of Fredericksen's honeys are certainly assertive; clover has the roasted notes of caramel, and for all the delicacy of the name *savory marsh flower,* the honey offers a grassy whiff—though I prefer Fredericksen's more lyrical description that it tastes "like a field after the bailing machine's gone through." Basswood is a widespread tree in this open land, but Fredericksen has found remote stands of ancient big woods and that's where he places his hives, marking them, for example, Hive 14 A at Painters' Creek. The flavor of that honey is delicate, herbal, haunting, there and not there and yet giving a clear message. Not every flavor needs to be superemphatic. This is the amount of flavor that one hive under a stand of flowering trees in central Minnesota can produce.

Staying at this level of production, eager to offer this type of revelation—that takes single-minded commitment. Fredericksen almost aligns his metabolism with that of the bees. He doesn't sleep through the winter, but he takes the season off, putting his energies into getting his independent-minded sled dogs to understand the notion of keeping to established tracks. When traveling, he keeps an eye on the hollows of trees, imagining the bee colonies that might live there. When I ask him how he'd like to develop, he sounds excited. He'd like to start working with minuscule four-inch honeycombs the bees could fill in three or four days. "You'd need a great source of nectar," he says excitedly, "but it would be a great snapshot."

Fredericksen doesn't want to grow bigger; small is his ideal. But for other artisans, growth is fraught, full of mixed feelings. It induces true apprehension, too, not some publicly performed agonizing but a real pang about what a bigger next stage will mean. It was convictions about individuality and uniqueness that set individuals on the artisanal journey—success and the opportunities it tenders narrows the trajectory and brings it back to the self. Up until recently, it had been a streamlined journey. Artisans had to learn the craft, then find ways of selling it. All those fairs they attended, all those markets where they folded up the legs of their plastic tables either exhilarated at having sold more than predicted or despondent that they were returning with as much stock as they brought—those were part of a single trajectory to establish themselves.

Then they did. And the business grew. For many, this growth has been the first conflicted situation they've been in as food artisans. The flip side of having your identity tied to your craft is that alterations in one are reflected in the other. The small-batch jam maker who's been working in her home kitchen worries what it will mean if she rents commercial space (she'll probably have to stop writing the labels herself, too). The coffee roaster with a growing reputation won't be able to run green beans through her fingers to determine how much toast they need. The cheesemaker considering extra accounts knows he can't keep up hand-ladling the chèvre into the molds to give it the airy consistency he loves. The benefits of growth are clear—the cost is rendering the connection between artisans and the very hands whose acquired skills made them artisans in the first place.

That is far more real and far more affecting than the tired binary choice between "craft" production and "mass" production (I'm guessing here, the opposite pole to craft never seems to get a name). Also unclear is the role money has played in the artisanal story. In early medieval Europe, the farmer with an extra flitch of bacon existed barely above the level of barter, rarely visiting town, and, when he did, in the ringing imagery of the French historian Fernand Braudel, held to "pressing his nose against the shop window of the marketplace." Capital and credits were the very "levers" that created a transition from the enclosed tract of self-sufficiency to the open horizon of a market economy, a bigger world where an artisan could confidently take his place. To insist that things cannot change, that things must remain on the scale they started, that growth is by necessity negative misses the overarching meaning of the term *artisan*. Artisanship may have come to mean small production, but it has *always* represented independence. That applies to the stained-glass craftsman working in the apse of a cathedral, to the breeches-wearing coppersmith in colonial Philadelphia, and to today's jam maker who in the dawn light loads the back of her Subaru hatchback with crates of jam-filled Ball jars to sell at the farmers' market. And it applies to the baker who has people lining up at 4 P.M. at Guerrero and Eighteenth Street to get a loaf of his fabulous bread.

HESITATION ABOUT MOVING TOWARD A new scale of artisanship is understandable. Artisans have to count on their existing customers to shift their own thinking so a business (even the word is iffy) can be considered valuable—for providing career paths, benefits, health insurance, and a future to its team. Launching a venture will always be emotional. The owner of Dogfish Head Brewery, Sam Calagione, recalls the feelings that came over him as he stood on a ladder outside the coastal Delaware business. "Behind the sign was yet another, older sign from the business that occupied the space before that last business. I asked Mariah not to take a picture of us removing that sign. As I pried it off the building, a flood of emotions came over me: Oh, my God! What am I doing?" All artisans who have reached that point have had their versions of that moment where dread, fear, and pride form a tight knot in the pit of the stomach. It's fitting to acknowledge those moments. That was the start. But to be hesitant about the size a successful venture grows to corrodes matters, strips meaning from words so a business that's achieved something special sounds like one more entity making claims about doing something or other slowly by hand. Ultimately, it doesn't let the sense of pride go the distance.

When I went to San Francisco, the deal between Tartine and Blue Bottle still hadn't been called off. In fact, Robertson and Blue Bottle founder James Freeman were booked to give an early November talk at a Silicon Valley speakers' series. That presentation would eventually be canceled, though by then the role I'd been cast in was clear: I was the supplicant journalist hoping for a break. My few days in San Francisco were spent both obsessively checking my e-mail to see if the interview I'd requested might actually occur and in a fugue of long walks around the city envisioning what the joint company might be. At Tartine I wondered what I'd do if I spotted Robertson—the artisan at the bench. Would I put aside the sense of being avoided and say anything? But I didn't see him. I wasn't able to find the Blue Bottle location in the Twitter Building but, farther down Market Street, in the Ferry Building Marketplace, amid the olive oil stalls, the salt tastings, and the offers of melted raclette, I watched lines of people waiting for an espresso shot or a pour-over inching toward the company's two counters. I drank my coffee elsewhere, at the Mill on Divisadero Street, a still-small-scale partnership between Four Barrel Coffee and Josey Baker whose pan-loaf breads make sensational toast. They prepared each slice to order at one of four pop-up toasters in the back, across from where a Clash record played on a turntable. I ate my buttered and cinnamon-dusted bread and wrote at the communal table and I was reminded of what a great San Francisco coffee shop can be.

The ease with which I slipped into that setting propelled me to the final Blue Bottle location I visited. After all, I entered coffee shops like a second skin, knowing their rituals of space, alert silence, and concentration coexisting with conversation and cluttered tables. But I didn't know what this new entity—coffee and great bread in incredibly minimalist settings—was aiming for. It seemed at once the opposite of the neighborhood spot with its left-behind newspaper sections and raggedy row of paperbacks, yet animated by it. I wandered down to the Heath Ceramics tile factory south of Mission Street on Harrison. It's a large building, with loading docks and the scale of industry—behind a glass window one can see the large factory floor. I've eaten off Heath plates many times in restaurants and it was nice to stand there watching where some of the low-key, high-quality items Heath makes are produced.

It was late in the day and there weren't that many workers inside, but the Blue Bottle kiosk had a dozen or so people sitting around it, talking or working on their laptops. The young man who made my pour-over was pleasant and explained that Tartine would occupy the large space on the Alabama Street side of the plant. I went around the corner to look at it. A construction worker was using an electric saw. He looked up from his sawing, decided I was harmless, and went on with his work.

This space would soon be filled with equipment and shelves, an ice cream counter, a pickling department, room for workshops, and the noise of customers. In our brief phone conversation, Robertson had sounded excited by the technological challenge of making more bread. He'd found mixers that worked so slowly they almost replicated hand kneading. "We'll catalog everything, even how many revolutions per minute they turn," he'd said with a real enthusiasm. And that was the tone I'd still hear when I circled back after the Blue Bottle deal was called off and I finally reached him. We spoke briefly—he was in an airport in Arizona, where he'd gone to check out portable and affordable milling equipment that replicated a millstone's burr without the actual millstone. Some big players had come forward, allowing him to keep planning expansion and thinking at a structural level. "It's easy to look busy running around opening and closing windows and worrying about humidity," he said. "But if you have a temperature-controlled room, you've just bought three hours.

Go read a book, go play with your kid, go grind some flour that will make the bread more nutritious and delicious."

After we hung up it struck me that he'd sounded like someone who'd worked something through. The baker who'd mixed dough in a wooden trough now understood that technology was the means of ensuring excellence. At culinary school he'd gone off the culinary path to get into bread. In Point Reyes, he and Liz had used bungee cords to fasten bags of the darkly singed bread and strawberry galettes inside the van they drove to the Berkeley farmers' market. Moving to San Francisco, the couple had given up the wood-fired oven and lost nothing in the process. Here was the next challenge. By minimizing the mystique of the artisan, he was drawing out that core paradox that an artisan food reaches completion when—dropped in a bag, foaming in a stein, wrapped in waxed paper or as part of a single-source sample pack—an item is handed from one person to another as a transaction. Leaving "small" behind as a measure of quality is hard, and standing by the back of the Heath factory I'd understood that it also involved a leave-taking. Tartine's path would lead away from the Northern California mists in which it had been formed. It was late by the time I turned away from the construction site. The fog was starting to roll down from Twin Peaks, pushing a rippling light before it across the city's roofs. Having this next phase in the hands of someone who'd waited twenty years to increase production to more than 250 loaves a day made me feel good.

TO HARVEST

The oranges that you see peeking out above garden fences and at intersections in the San Fernando and San Gabriel Valleys that flank Los Angeles are mementos from when the area was full of orchards. These orchards, already ranging in long straight lines by 1920, endured the days of oil fields being drilled and film studios being built only to succumb in the post-war housing boom. Fueled by the aerospace industry—many headquartered in the area—home construction exploded, and the orchards made way for subdivisions of identical houses with enclosed yards, each one shaded by a single spared tree. In certain parts of the San Fernando Valley—that vast swath of land stretching from the poolsides of Encino to the chaparral-studded hills of Pacoima—the trees seem to be waving to one another, their loads of fruit like bright flares in the Valley's flat white light.

It's Food Forward's mission to not let such fruit go to waste. The group's founder, Rick Nahmias, calls the San Fernando Valley "a decommissioned citrus orchard," and since he started the group in 2009 with a simple backyard pick it has grown into an energized, urban food-gleaning organization. Food Forward runs a switchboard of sorts where donors call to get their trees picked by a team of volunteers (who get the fruit immediately to a food bank). It also mobilizes teams at seventeen weekly farmers' markets, consolidating all produce that farmers don't want to drive back with them. Food Forward's twelve-week Can-It Academy gives in-depth training to a constituency of food preservers with diverse interests, from preserving the bounty of their kids' edible schoolyards to starting a home-based jam-making or pickling business.

Since early 2014 Food Forward has also been attempting to forage on the same scale as industrialized food operates. The purchase of a twenty-four-foot-long truck allowed it to send a driver down to the L.A. wholesale food market every morning before dawn. Nahmias understood this was a necessary step. "Produce comes in a direct shot from the Central Valley, from Asia, Mexico, and Riverside County," he says. Borrowing from Zola, who described Les Halles, the central market in Paris, as "the belly of Paris," Nahmias calls the massive sprawl of warehouses and idling eighteen-wheelers "the biggest belly in the country." He estimates there are 2.4 million food-insecure people in L.A. and adjoining counties, which makes being involved with this huge source of produce all the more important. "We thought we'd do a couple of hundred thousand pounds," he says of the first year's wholesale gleaning. "We ended up distributing four million pounds in the first year." He is a man who can roll off figures easily because they represent the value of the shared work, but there are certain ones that he gives with obvious misgivings. Salvaging four million pounds means an awful lot more went to waste. "There's joy in the harvest," he says, "and anger at the waste."

Waste is the underside of everything that's uplifting about

food. We're much more comfortable with the idea of frugality. Thrift—that's a value we can trace from the repurposed New England boiled dinner that becomes a red flannel hash, to the tenant farmers of James Agee's *Let Us Now Praise Famous Men* busy canning produce, to guerrilla fermenter Sandor Katz, whose bread recipes call for using all manner of leftovers. It even reaches to me. I would watch my grandmother, working alone in her galley kitchen, run her thumb inside the shell of a freshly cracked egg to ensure no egg white was left behind. Waste can be just as personal, but it doesn't inspire; it hurts. In one of the most affecting passages of Roy Choi's memoir, *L.A. Son,* the star chef writes of the culling he'd perform when Silver Garden, the restaurant his parents opened in Anaheim in 1978, ceased being a success. "My mom continued to prep as if three hundred covers were coming in each night, stuffing plastic bucket after plastic bucket with marinating meats and her fantastic kimchi. Night after night, these dishes just sat row after row, orphaned and waiting for people who never showed up. Some of the food rotted and decomposed in the back of the walk-in. Sometimes I went in there, quietly, cleaned it all out."

That *quietly* is powerful. We're all aware of the sense of unease we get when throwing out something that was once edible. But we're busy and it's easily rationalized. Who can keep up with the tomatoes at certain times of year? It takes the kind of conscience most of us are not graced with to do things differently. Or else it takes a shock. In January 2009, Nahmias was in a dark place. In July of the previous year, he and his longtime partner, Steve, a psychotherapist, had married a few weeks after the State of California started issuing licenses to same-sex couples. It was a brief window. Already active locally in the Obama campaign, he had started working for the defeat of Proposition 8 once it became apparent that the measure, which would have made his marriage invalid and same-sex marriage unconstitutional, would be on the same November ballot. On election night, Rick and Steve gathered at what they thought was going to be a double-victory celebration at a Hollywood theater only to realize they were living in a sort of split screen. There was the jubilation of seeing the newly elected first couple dancing onstage in Chicago and, in their own state and among the friends who had gathered with them, the gnawing realization Prop 8 was passing. "For any gay person," Rick recalls, "it was like being hugged and slapped at the same time."

Instead of celebrating, he and Steve drove home in stunned silence. Over the next two months, Rick developed an uneasy form of detachment from his life as a commercial photographer. "How do you get your mind around the fact that 53 percent of the population believe you don't deserve a basic right?" he remembers thinking. By then the country was also ensnared in the sharpest economic downturn since the Great Depression. The San Fernando Valley, where he and Steve live, was hit particularly hard, especially its northern rim, by housing foreclosures. For Sale signs creaked in the wind that blows down through the mountain passes and runs along the cinder-block garden walls. With the help of the community-organizing group OneLA, activist priests like Father John Lasseigne of Mary Immaculate Church in Pacoima organized hundreds of congregants to negotiate as one with the banks. For many others it was too late—the shrinking economy had become about hunger. "There were people lined up outside food banks two miles from my house that I didn't even know were there," Rick says.

It was in that frame of mind that one day he took his dog, a rescue mutt named Scout, for a walk. At twelve years old, Scout walked slowly, and, as Rick's mind wandered, he noticed the fruit-laden tangerine and navel tree in the garden of his neighbor Heather. Rick is a son of the Valley. Growing up in Tarzana in the late '60s and early '70s, he and his friends would gather in the golf courses that abutted several developments and use the greens and sand traps as their playgrounds. It was a childhood, as he puts it, spent "in the bubble of carpools and suburban sprawl." Riding in the backward-facing seats of the family's wood-paneled Esquire wagon, he'd seen many, many orange trees go by. And he knew what happened. A good-faith effort was made to pick some, then everyone got tired of eating them, and the living-off-the-land lark was allowed to run its course. The fruit either rotted on the branch or eventually dropped into the lawn and decomposed.

Here was food, he thought. Two miles away people were lining up for food donations. Had he lost such agency in his life that he would just let the progression of things play out the way it always had? He was tempted—sorely tempted—to do nothing. Marriage could just be snatched from people, why fight anything? But Scout's pace kept him there, pacing slowly along the garden's edge, thinking. *Would it take so much effort to pick that fruit and donate it?* By the time he got home, he called Heather, the neighbor on whose property the trees stood. As fellow dog owners they'd grown friendly, and Rick or Steve would sometimes watch her young daughter when she was stuck for a babysitter. She told him to go ahead; she'd already picked all she could possibly use. Not quite sure how to proceed after his grand announcement, he put an ad on Craigslist, announcing the pick that weekend. Two young women showed up on Saturday morning. One was an experienced enough urban forager that she'd thought to bring banana boxes and empty berry flats she'd retrieved from a supermarket Dumpster to put the fruit in.

It took three weekends to pick the trees clean. By then, the director of the SOVA food bank on Vanowen and Hayvenhurst had grown so used to seeing Rick and his helpers unload boxes from the back of Rick's VW Jetta that he told him if he was interested in picking more, he had just the orchard for him.

The Valley has its own form of culture. A squat run-down building in Van Nuys might turn out to be the recording studio from where Nirvana's grunge anthem "Smells Like Teen Spirit" went out into the world. Along with news of homecoming night and exam schedules, the towering sign outside North Hollywood High proudly announces it's the alma mater of Susan Sontag and Cuba Gooding Jr. Rick is embracing his local roots when he describes the orchard's location in Valley terms. "It's where Winnetka dead-ends at Devonshire," he says, "on what used to be Lucy and Desi's country estate."

Back he went to Craigslist, also putting up flyers on the surrounding Starbucks community boards announcing a pick and a potluck. By the end of that March day, the team had picked two hundred boxes of Valencia oranges and three hundred pounds of grapefruits. Even more surprising for Rick, twelve of the people who'd climbed the ladders, plucked—twisted, really, so as not to tear the peel—and hauled box after box had taken the time before departing to leave him with their phone numbers and e-mails and wanted to be contacted if he ever did it again.

Rick continued with this small model of food activism, but what he ended up with was something much bigger. The Los Angeles County Agricultural Bureau estimates there are one million fruit-bearing trees in the county. Plenty of trees to pick. In a way he had found not only what would become Food Forward's mission but also the method by which it would work. The operation is nothing if not efficient. The group is responsible for the intelligence gathering on possible harvesting spots and supplying equipment and the stacks of flat boxes imprinted with the logo. The people who sign up for a pick supply the sweat. What they receive in exchange is the kind of jolt of positive energy that is the coin in which volunteering is paid.

To even dip into the volunteering opportunities the organization offers is to feel that energy oneself. When I signed up to do a pick on a weekday in Lake Balboa, one of many suburban neighborhoods in the Valley, I was sent a confirmation e-mail with the time, address, and length of time I should expect to stay. On the day of the pick, as I drove down the street, I saw a Food Forward sign had already been fixed into the lawn of the ranch house—like a campaign device by an orange-faced candidate—to guide volunteers and inform passersby. A young woman was our pick leader; an older woman in her sixties who'd taken early retirement completed our trio. Within minutes we were easing ourselves in through the gate that by arrangement had been left unlocked by the homeowner and entering the yard.

It was a peculiar feeling. A dog's chew toy and a child's tricycle lay in the grass. Someone smoked, though the ashtray was kept outside. For a moment I felt like an intruder. Within minutes I had a picker's bag strapped to my shoulder, a pole with a clawlike end in my hand, and had internalized what might as well be the group's motto: *Twist, don't pull.* I was amazed at how seriously I was taking it, struggling to get those oranges that reached across the fence and over the neighbor's pool. A few minutes before the two hours were up we'd raked the area under the tree clean ("Always good to leave it better than you found it," our pick leader said), we'd loaded the pickup, and I saw 150 pounds of oranges driven off to a food pantry.

The market-gleaning teams that Food Forward inaugurated in 2012 have since extended their activities into a realm of food more public than the backyard. Since 1981 the Santa Monica farmers' market has taken place every Wednesday morning on several blocked-off streets a few blocks from the Pacific. It's a great gathering of dusty pickups, chefs with trolleys, and shoppers there for a quality of fruit and vegetable no supermarket can provide. There's always time to pause and chat with acquaintances, but in the early hours every shopper is focused on something important, like getting to the rancher who brings loganberries for about two weeks a year before his trees are done. "Giving Good Weight," John McPhee's 1978 *New Yorker* piece on the experience of working behind the stalls at the Harlem, Brooklyn, and Upper West Side markets, brings the physical aspects of urban provisioning to life. "You people come into the market," he writes about his time behind the stalls, "and you slit the tomatoes with your fingernails. With your thumbs, you excavate the cheese. You choose the string beans one at a time." The level of mistrust may have lessened (and the organic status inquiries shot up), but neither the forceful melon thumping that goes on in Santa Monica nor the prudent pressing of the pluots (to say nothing of the outright grumbling about the price of the blue lakes) can dampen the murmur of satisfaction city dwellers make before great produce.

That first pick was an example of Food Forward's ability to invent a mechanism practically on the spot. Sarah Spitz was there that first day. "We knew nothing," she recalls. "How are you going to track donations, weigh food, what are the organizations that you're going to give it to? One thing we figured out quickly is don't ask for donations until almost the end. Don't interrupt the farmers while they're trying to sell." Today those principles are applied in the seventeen local markets Food Forward gleans in. The one in Torrance, in L.A.'s South Bay where my daughter Isabel and I volunteer at one weekday morning, is run with their signature efficiency. The e-mail told us to meet behind the kettle corn stand at the market's edge. Two volunteers in sun visors sign us in and give us a basket of name tags of previous volunteers to go through (to save paper). Meanwhile, Tom, an older man who participates every week, goes for the weighing scale and trolley he keeps by arrangement with the custodian of the nearby rec room. The names Isabel and I go through represent a nice cross section of L.A. (Rowena, Hyung Joo, Armando, Agnes, Brittany, Zoe), but not finding our own, we write out identifying tags.

We move between the stalls, picking our moments to politely ask farmers if they'll be donating this week. "Yeah, give me three boxes," a man with a table heaped with cucumbers says. We assemble the boxes and write the name of his farm on the side. We'll be back in half an hour to pick it up. "Don't ask that one," Tom says, not intimating the person is mean, just that they're not into donating their stuff. Cool. Unlike the Hollywood or Santa Monica markets where a chef might be trundling through with his hand truck, Torrance's market is about good staples for locals. A line has formed in front of the stand selling hormone-free eggs. People bring their own empty cartons. The two guys selling musubi, that Hawaiian energy bomb of Spam in sticky rice wrapped in nori, have just sold out. A half hour later, as the tables are starting to be folded up, we walk back around gathering the now full cardboard boxes, ferrying them back to our post, where we weigh them, make note of who donated, and divide the produce among community groups, some of whom have already driven up.

The driver from a retirement home stacks boxes with speckled greens into his pickup. He also got a lot of thyme and basil and a box of thick, ruddy carrots just the way I like them. For a moment, I'm tempted to pocket some except I figure that's being a very bad volunteer. Instead, I help a young woman cram boxes of eggplants and fresh stone fruit into the back of her Nissan. She's driven over from El Nido, a low-income child services organization for pregnant women, toddlers, and special needs kids housed in a two-story stucco building on Manchester Avenue in South L.A. I know the building; its little Astroturf playground is a splash of vivid color under the jets coming in to land at LAX. I'm glad I'm loading boxes of peaches and cucumbers into the trunk of her car. In that landscape with few trees, little shade, and much heat-trapping concrete, these peaches and cucumbers seem like something vital and alive. It isn't until she's driven off and I realize I've helped them get boxes of great produce that I feel the full Food Forward rush.

"A lot of our new volunteers get what we call 'fruit goggles,'" Rick says as he gives me a tour of Food Forward's North Hollywood headquarters. "Once they get the idea of saving fruit they see it everywhere." I nod because I know what he's talking about. Somewhere in the glove box of my car is an index card with the scrawled-down cross-streets of two intersections where I've seen loaded branches overhang the sidewalk.

Food Forward's headquarters is located in a squat industrial building, where people are busily working on laptops and computers. The company logo, modeled on an old citrus box, adorns the reclaimed wood walls. The key piece of equipment, the metal clawlike picker that you can shoot up a pole into the thicket of leaves to grab fruit beyond even the highest ladder, serves as sconces over bare bulbs.

There are crops other than oranges in the Valley, and as soon as Food Forward volunteers started picking Meyer lemons, they had a problem. The Chinese sweet citrus has a soft skin, which makes it impossible to donate to food pantries because it easily rots. Perhaps they could learn to make jams and sell jars for fundraising. Rick had recently met Ernest Miller, a well-regarded local chef, and they agreed that Food Forward would sponsor a twelve-week course that would lead to master food preserver certification. In 2011, Miller had been instrumental in resuscitating the food preservation program the University of California Cooperative Extension had started in 1914 to teach homeowners proper canning techniques. Though it had thrived during the Depression and the era of Victory gardens of World War II, it had all but ceased operating by the 1990s.

When I catch up with Miller, he is leading one of the seminars at the top floor of the Santa Monica Promenade, a recently remodeled mall by the Third Street pedestrian area. Professorial, though with a quick laugh, he stands in a white chef's coat before eighteen students. Some of them come from the thriving edible schoolyards movement, which plants gardens maintained by students. Two young women want to start their own business, taking advantage of California's recently passed cottage food laws, which allow certain foods—particularly jams—to be sold when made in a home kitchen.

It's a warm night and most have opted for jeans and T-shirts, a good option when working over kettles of steaming water. The scene feels completely contemporary, but the questions directed to Miller are eternal ones. Why is my jam not coming out? How do I can beans? How do I know when they're ready? How can I be sure I'm not poisoning people? Miller is thorough in all his answers. He never makes any technique sound easier than it is. He is trying to make dreams viable reality by grounding them in food safety science. "Let the bugs do their thing," he says with a certain nonchalance about how chiles and onions, salted and kept in a crock, will eventually produce a lacto-fermented hot sauce. His dramatically timed delivery can also get the students scribbling in their notebooks fast. "Botulism spores are everywhere. To become active they need a low-acid, anaerobic environment." Pause. "Like a sealed jar." With that he points at a slightly dinged large pot with plastic handles and a sealed top sprouting a dial gauge to read the pressure inside. "Green beans are not acidic enough for safe home water-bath canning, so we use the pressure canner, which raises the temperature of boiling water from 212 degrees at sea level to 240 and kills the spore." After a short break, everyone is stuffing jars with carrot batons and taking turns carefully placing them in a hot water bath with rubber-tipped, insulated tongs. The process is being timed on a cell phone, and the students gather around a high, steaming stock pot to watch. "If you see little bubbles, it means the vacuum is being created," says Miller. "That's good."

BY EARLY IN 2014, ONLY two years after its first farmers' market effort, Food Forward had developed a reputation for its efficiency. "We were getting random calls from people who found our name on the Internet and wanted us to pick up six pallets of squash in the downtown market. So we became an ad hoc recovery service. We purposely didn't have refrigeration because we didn't want to hold anything. We're the transfer point between donor and receiving agencies." With more and more calls coming in about entire pallets, Rick understood he had to start foraging at the downtown market, a vast spread that stretches along Olympic Boulevard between downtown and the Los Angeles River. Rick wasn't quite sure how to engage with the minicity. The answer came in the form of a six-foot-two ponytailed musician who had transferred from punk to opera, Luis Yepiz. Luis started to drive a truck down to the market early in 2014.

I meet Luis before dawn at 4:30 A.M. one day to accompany him on his market run. He's already driven from his home in El Monte to pick up the truck at the MEND food bank in Pacoima. On any index of poverty in Los Angeles County, this area ranks high. In obesity and soda consumption, it measures alongside South Los Angeles among the highest areas in the state. In a map UCLA health researchers published that represented the likelihood of limb amputation due to diabetes in shades of blue, the northern Valley, like South Los Angeles again, is the deepest blue. It can't all be pegged to diet, but lack of fresh produce in markets is what Dr. Dylan Roby, coauthor of the study, calls "a social determinant of health." Compounded with other factors, health care becomes "a matter of tertiary intervention. You deal with disease when it's a crisis." When Food Forward got involved in recovery from the downtown market, it entered a new phase, perhaps one that had been awaiting it, operating within the inescapable dual narratives of modern American life: waste and hunger.

Rick had prepared me for Yepiz's big personality. "He can look a guy in the eye who has just donated twenty thousand pounds of produce and say 'I can't take that' about a pallet he doesn't like." Luis's technique is to get noticed—and not be shy about asking. No sooner than we'd parked, he was scurrying up loading docks, charging through industrial-weight fly flaps with a *"Que pasó, raza"* to the workers who might be taking a predawn break. He's been doing it long enough—six months now—that he knows who to bother talking to, who communicates with a nod. It's a world where he has to encourage generous impulses yet also watch out for brokers who would love to get rid of rotten produce and save themselves the forty-five dollars a ton they'd have to pay to have it hauled off as waste. There is a lot of that. I separate myself from Luis while he's negotiating in the depths of a refrigerated warehouse and see a young man emptying crate after crate of melons into a Dumpster while carefully folding the cardboard boxes they come in. The packaging still has value.

By eight in the morning, sunlight has broken, and the noise of traffic from the rest of the city has reached the speed the wholesale market has been operating on all night. There's one last stop, a particularly good donor, and we're standing on the dock when a warehouse man wheels out a forklift with a massively high stacked pallet of peaches. He knows better than to drive it into Luis's truck. Luis reaches high and takes one peach from the top and takes a bite and nods. Still, the driver knows not to drive it on. Luis reaches into the bottom third and takes another. Again he bites, chews, nods. "You have to be prepared to eat a lot of crap at this job," he says. He doesn't need to nod for the forklift to push it onto the last empty spot in the twenty-four-foot-long vehicle.

Before noon, Yepiz has driven to Heart of Compassion, a large volunteer-run distribution center in Montebello. Guys with shaved heads unload some of the pallets and we serve ourselves a quick early lunch of salad (from a reclaimed box) and stew (also with salvaged bell peppers), off trays they've put out on a trestle table. Then we head out across town to the Dream Center, a transitional housing center located in the former Queen of the Angels Hospital off Alvarado Street in East Hollywood. Built in 1926, it's an elegant seven-story tiled structure in the Spanish Revival style. A group of experienced residents swing into action when we arrive, transferring the pallets from the truck into the refrigerated containers they have running on generators in the parking lot. Later, they will transfer some to their own kitchens and subdivide much of the food into individual bags they donate to people who line up outside every morning.

In one morning, I've learned to dodge forklifts, but I know that I'll only be in the way if I try to help move things on the truck. I walk around to the front of the building, where I stand looking down at the 101 Freeway. Traffic pools in the direction of downtown. In the distance, a helicopter chops through the pale haze. This is a different city from the one I inhabit. Restaurants are my beat. In that world, seating is an index of power. The other night I heard a maître d' say over the phone, "Of course you can have the VIP table, it's Monday." And I don't think he realized how he'd just cut the guy down. This full truck measures something else. L.A. is a city where eating food is public. There's the wood rail that runs around the parking lot at Tommy's Burgers on Rampart where people perch to eat chili fries. Dusk settles over groups ordering from the taco stands that pop up under strung-up lightbulbs near MacArthur Park. A parked pickup might be full of watermelons or mounds of ripe plantains ready to be bought and taken home and fried. On Vermont

Avenue I often see women selling *flor de loroco,* a white flower that's folded into the masa of Salvadoran *pupusas,* lending a sharp bite to the cheese filling. It seems abundant, lush to see all those white flowers, but it's just another mask over the face of hunger in the city, a sensation that hits me now, vague and inchoate, as an unseen pang. "That's it!" shouts Luis with typical pep when I rejoin him. He rolls down the back door of the truck, and we hop into the cabin. By the time we've arrived back in Pacoima, it's one in the afternoon. When the food bank staff here have unloaded the truck's last pallets, we've distributed fifteen thousand pounds of food. Luis will start again tomorrow at four A.M.

HOW DOES THIS TRANSLATE TO the private realm? Being even slightly involved with Food Forward has had a real effect on the way I eat. There's the practical matter of finishing everything on my plate that comes into conflict with my role as a critic, which often involves ordering an extra dish or two. When I cook, I cut off the carrot top as precisely as possible; wash instead of peel the cucumber. I make sure to eat the cheeks and the belly meat on the branzino. It's not flavor but more an awareness of waste that links us to an older understanding of food. Putting by, pickling, those techniques speak to a sense of self-reliance that is very different from being a docile consumer. I wouldn't call it "ethical eating." (When I was a too-serious sixteen-year-old using a wood spoon to eat brown rice from a bowl I'd carved, I would have merited that description.) And it's not exactly a flavor that's gained, but by being aware of waste and seeking to avoid it, we find a link, perhaps even stronger than any recipe, to the generations that put away, that strung out a ham through the winter, that made sure every last piebald bean went into the basket. American cooking seems most resonant when it's reusing the Thanksgiving turkey bones to make the next-day soup.

With its logistical resources, Food Forward can now take on any size orchard—even the last few remaining large ones that somehow avoided being uprooted. The volunteers will go through the grove at E. Waldo Ward, a small company that's been making jams in the canyons that reach into the San Gabriel Mountains in Sierra Madre for more than a century. The orchard that skirts the driveway of the Huntington Gardens in San Marino was protected by the estate. The four hundred trees at the center of Cal State Northridge were protected by Dr. Robert Gohstand, professor of geography who in the late '80s got wind that the grove which had stood as the university was built around it in the '50s was going to be dug up.

The son of émigrés from Russia to Shanghai, where he'd been born, and then to California, Dr. Gohstand was more interested in tracing the history and meaning of Moscow's underground—of which he was a specialist—than the Valley's agriculture. But hearing the grove was slated to be uprooted seemed to him to be denying recent history, a fact that he put into a letter to the Facilities Planning Commission. "The suggestion that some orange trees be scattered elsewhere on campus as a sort of sop to nostalgia for the area's agricultural past is not at all satisfactory to me." Entering with some glee into the reasoning that it would be tearing up what would then be honored, he argued that preservation of the nine-acre grove itself was a commemoration, "something no arch, column, statue, building or, the Lord knows, parking lot can hope to do."

Every year on a Sunday in April, Food Forward picks these oranges, setting out boxes and getting trucks deep into the grove along the few paths. By ten A.M., when I get out there, the groups have already been busy. There's a large contingent of volunteers from Bank of America wearing red T-shirts and an equal amount of men and women from an organization called Gay for Good, climbing up trees, organizing themselves between those who want to be on ladders and those who want to haul boxes to consolidating points. Wearing shorts, sneakers, and a big smile, Rick is busy at the center of it all. "The Seabees will come by later," he says, giving the shorthand for members of the Naval Construction Force who are driving from the base at Port Hueneme for the evening's last sweep.

I have a notebook, but of course I'm soon hauling boxes. What the effort speaks to is a form of decency. We're not the figures in the Millet painting picking our way through the stalks. We're the gleaners of today, who apply a little zinc oxide to our noses before heading out, who arrange to leave the backyard open, store the trolley with the park manager, meet by the kettle corn stand, hear the satisfying pop of the Ball jar sealing, and, when jolted from our daydreams, become part of a chain of people handing boxes full of oranges toward a truck where they're stacked and driven away.

TO TELL

Taco María occupies a corner by the entrance of Costa Mesa's OC Mix, a rustic-mod cluster of businesses off the 405 Freeway in Orange County, California. The openly displayed coffee roaster and the bins of vinyl records may be a bit affected, but there's nothing wrong with the message it's sending, that this is the antidote to the sameness of commercial malls. With its white tile interior, dark wood accents, and chefs wearing selvage denim shirts, Carlos Salgado's restaurant, Taco María, fits right in. I like to come down here—it's an hour from L.A. if I beat traffic—and sit at the counter, eat, and watch the cooks work.

Tonight a young woman wearing a bandanna tends the grill. She turns cactus paddles and ribeye cap over the citrus wood embers. "Yes, Chef," she answers when Carlos calls out an order from the other end of the counter. The tickets lie in front of him, held down by a foot-long piece of dark, flat wood. He wears small, black-rimmed glasses and what looks like a gray work shirt, though made of finer cotton. A few other cooks pivot between counter and stove in the twenty-foot-long open kitchen. They all have their own compact, slightly elevated wooden cutting boards; they hand elements of dishes to one another—a fillet of striped bass dusted with achiote—working quickly to not lose the heat. The cook in front of me finishes a bowl of *chochoyotes,* shelling beans and tiny masa dumplings, with a few drops of hot oil that open all the flavors up.

Carlos doesn't serve dessert at Taco María. The space is impossibly tight—from my counter seat I can see the dishwashing station crammed into the tiny back kitchen. He doesn't want to take even inches from the compact dining room with three tables or the terrace, the restaurant's largest seating area that looks out over a little garden where low-slung Adirondack chairs are scattered invitingly around a fire pit. Tonight he's fashioned a little something, though. Yesterday, at the end of the dinner service, one of his cooks scattered the grill's embers over a pan of fresh berries from J.J.'s Lone Daughter Ranch. The heat seared them and brought out an earthy quality in the fruit as it cooled.

This morning the berries were cleaned, pureed, and blended into masa, the corn building block of tortillas. Diluted into a drink and flavored like this, masa becomes *atole de elote,* a warm drink that sidewalk vendors ladle from insulated thermoses. Carlos keeps it in a bowl by the grill. He hands me a speckled cup of the thick, faintly purple liquid. It's nurturing enough that just wrapping my hands around the cup feels comforting. Cooking the berries muted their freshness, but in such a way that they harmonize excellently with the ground corn.

Carlos's menus are full of those kinds of oblique references. They are allusions to home cooking as experienced by a first-generation Mexican American in Southern California who just also happens to have been voted one of America's best new chefs by *Food & Wine* magazine. His dishes elevate without condescending, so the *chochoyotes* may recall a bowl of pinto beans, and the sliver of red onion steeped in vinegar and hibiscus he might toss—okay, tweezer—onto a lunchtime chicken *mole* taco is a garnish from a street stand. His version of *atole* may not be the expected dessert in an ambitious modern restaurant like this where pale, dry Albariño is served in tumblers, but it's a perfect way to end a meal.

The reason I keep coming down here, looping through freeways—driving south on the 5 through Norwalk and Buena Park, switching to the 55, that freeway that runs down through Orange County before jogging onto the 405 in Costa Mesa—is that in Carlos's fascination with corn I glimpse something metaphorical, a powerful statement about modern dining. Despite so many changes—the open kitchen as stage for one—chefs are often still painted as artists whose masterworks are recipes. That's a vestige from a time when a chef was judged on how well he glazed the Dugléré over the paupiettes of sole, evaluated on how long the crust remained crisp around the beef Wellington. Today's chef may find inspiration when she holds a peach in the farmers' market or when he runs uncooked Anson Mills stone-ground grits through his fingers. Heritage is using produce that jostled to market on the back of a pickup, a moment of awe before the rippling flanks of a Yukon king salmon, or performing a technique the way a mentor taught. For chefs today, a recipe is not some ritualized set piece but how they get to transmit what speaks to them.

I try to remain aware of that thread amid all the bustle of dining. When Ludo Lefebvre grills Iberico pork in the tiny Trois Mec restaurant behind an L.A. gas station, is he channeling *jambon au Chablis* from his native Burgundy? When Suzanne Goin uses arugula, the rib always snaps with a peppery jolt; it's a statement.

Watching videos online, I see a deep connection to an intuitive preparation in the face of Sean Brock as he pours cornmeal mush into a heated skillet to make his corn bread. I hear it in the voice of Marrakech-born Mourad Lahlou, who draws on the Moroccan tradition of salt-cured lemons, grating barely brined peel over sashimi, using the pulp of others that have cured for a year to lend umami to a braise. "I can't overstate the significance of bacon dashi to us at Momofuku," David Chang writes of a building block broth in his cooking. "It's not the dashi itself—though it is delicious—but the thought process that went into it. The successful transposition of bacon from Tennessee for Japanese dried and smoked fish was a driving inspiration of how we cook." Sometimes, to get the element that speaks, you have to make it yourself.

Using an ingredient that is at the root of a heritage is even more compelling. The corn Carlos uses are ancient heirloom varieties. It's dry when he receives the big burlap bags, and he cooks the kernels in water and powdered slaked lime, or *cal.* The name for that process is *nixtamalization,* the chemical reaction that softens the outer skin of corn and makes its many elements nutritionally available. Without that step, ancient peoples were racked by pellagra, a skin-corroding malady; because of it the corn crop became the building block of Aztec civilization.

Of course, at one stage Carlos wasn't thinking of corn as anything so ancestral. Corn? That was the domain of the quickly grabbed tortilla, or chips tossed in a pan with green tomatillo sauce and drizzled with sour cream for an order of chilaquiles. His concept of the ingredient was brought home to me one morning when I watched him grind the cooked corn into masa the restaurant would use throughout the day. After retrieving a large bucket of the corn that had cooked the night before, he drained it over the sink in a perforated pan. I looked at the kernels. Each one had a broad crown and a narrow stem, an elongated shape you don't see in today's hybrid varieties. Grown by farmers in *milpas,* or traditional checkerboard fields, it seemed about as far from a monoculture as you could get. He'd quickly assembled the parts of a five-horsepower mill; on went the hopper, the auger that pushed the corn forward, and two Frisbee-sized millstones carved, like the traditional *metate* concave grinding surface, from volcanic rock. He'd started to grind, intermittently splashing ice water over the kernels as they churned to get the masa to the right consistency.

But it's not just the texture or even the flavor he's after. When he flips one of the hand-pressed, deckle-edged tortillas on the griddle, he sees a chain of beneficiaries. There's the corn that finds a vital life in a modern setting, the customer who tastes the integrity of an ingredient, the farmer who's encouraged to keep growing it, and money that goes toward a local economy. This broader vision was a destination he had to reach, and he did it with the help of mentors such as Daniel Patterson, from San Francisco's Coi. Almost immediately after starting there in 2006, Carlos had been struck by the way the kitchen chefs crafted their own butter, culturing the cream, letting it mature, patiently draining off the buttermilk and using it for other purposes. The recipe had come from Soyoung Scanlan, cheesemaker at Andante Dairy in the town of Petaluma north of the city. She'd originally started making it for Thomas Keller at the French Laundry. Like Chang's bacon dashi, Patterson saw it as a gateway ingredient to an entire approach to cooking and, once letters reminiscing about sitting on the porch shaking a jar of cream to make butter started to arrive, of the resonance a single ingredient can contain. Carlos's journey to this meaningful place—at once intimate and sweeping—would require him leaving Orange County and, just as decisively, returning. What's a system of freeways to me is to him the grid of the past.

Recounting Carlos's life story even close to chronologically requires starting a half hour up the 55 Freeway in Orange, where in 1986 his parents, Maria and Gregorio, started a modest restaurant amid the body shops on Batavia Street. La Siesta is still there, and Maria and Gregorio still work at the counter, welcoming mechanics and city workers for an early-morning breakfast of eggs and nopales or a lunch of enchiladas, the sauced chicken-filled tortillas moist and filling. Customers might even sit at the table where Carlos did his homework together with his older sister, Silvia (Amy was then too young).

A culinary story is also a journey, rich in challenges and key moments, though the destination is not a place but a style. It draws on everything that's been experienced in the heat of kitchens but expresses it in techniques, flavors, ingredients, even the way of thinking of food. Every young cook we see—or do we even really see them?—going to a shift in houndstooth check pants and carrying a knife roll through a city is involved in that learning period. With each bus ride to and from work, with each night's dinner service, aspiring cooks' sensibilities are being sharpened, what they have to express becoming apparent, even if at first it's only to them. In some jobs you come for the information and leave; with others you resist that urge, choosing instead the discipline—and eventual knowledge—that repetition grants. At times, a cook even has to reach a breaking point for matters to become clear.

Carlos was just such a cook one morning in 2006 when he walked up a San Francisco hill toward Coi, knife roll in his backpack. He'd been working under chef Vernon Morales at Winterland, a Fillmore District restaurant that despite chowhound raves about its bacon ice cream and inventive cooking had been unable to attract enough diners in the city's tough scene. On their last night's service, after the final order had been put out, beer and bourbon had flowed as cooks and waiters sat up on counters, exchanging contacts and sharing in one another's company as a working crew one last time. When Carlos returned the next day to pick up the tools he'd forgotten in the midst of the previous night's jovial mood, he found Morales packing up boxes. Feeling responsible for a young cook he'd come to value, Morales told Carlos there was one restaurant he should apply to in the city—and it was Coi.

There had been a minuscule period of time when San Francisco hadn't known what to make of Patterson's serious yet irreverent restaurant. It was wildly ambitious—the culmination of his two previous spots, Babette, which he'd opened in Sonoma in 1994, and Elisabeth Daniel, which he ran from 2000 to 2004, but putting Coi amid the strip shows and video arcades of North Beach's Broadway seemed contrarian. Patterson pushed things further with a name that seemed intended to trip up the unknowing. Pronounced "kwa" (he'd gotten it from an online search), it was an ancient French word for calm, the mood that should reign over the twenty-seat lounge and the twenty-nine-seat dining room. The *San Francisco Chronicle*'s Michael Bauer had described the stakes that July. "Will the third time be a charm for Daniel Patterson," he'd written in his three-star review, "or is it three strikes, you're out?" A few weeks later, Carlos was walking inside, looking for a job.

He saw immediately how incongruous the setting was. There was a stripper taking a cigarette break in the alley the restaurant shared with a club. Carlos entered an already busy kitchen, the cooks working at their separate stations all aware—though not pausing in their prep—that someone like them had just walked in. "I found Daniel and asked if he had any positions open," Carlos recalls. "He said, 'We don't have a position but we have a good group of people.' That sounded good." He got sent to the pastry station and opened the knife roll he'd brought with him, which in addition to a set of knives and spatulas contained a refractometer for measuring sorbet syrups and his newest acquisition—a digital thermocouple thermometer capable of measuring tenths of degrees.

Technology and food had been wound together for Carlos ever since age fourteen, when a neighboring business to La Siesta had asked him to do its data input. The owners didn't have the patience for all the disks and drives; it was second nature to him. The world of command, prompt, and the blinking cursor was a type of release from the other one where he'd already been cast a fixed role: the smart immigrant kid. His mother is from Jalisco (home of tequila), a province that extends from Guadalajara to the coastal resort of Puerto Vallarta; his father is from Guerrero, farther south on the Pacific coast, known for seafood preparations like pescado Zihuatanejo. They had come to the United States separately and met while working at Alphie's, a coffee shop in Torrance, where Gregorio was kitchen manager and Maria was a waitress. They were focused on their children's education, and once they realized that the bilingual program in the Orange County schools had lower expectations for students, they insisted all three children attend regular classes (where Carlos was a consistent AP student) and become articulate in English as opposed to Spanish.

If that was a form of alienation, it was only one among many. In *Orange County: A Personal Story,* author Gustavo Arellano—like Carlos, born in 1979—paints a picture of existing on two levels: a *zapateado* folkloric dance may break out at home, but all you can think of is the latest version of Sega Genesis. Mexico was already a distant place. The neighborhood around Magnolia Avenue in Garden Grove where the Salgados lived was half Latino and half Asian. Carlos had as many Korean and Vietnamese friends as he had Hispanic ones, all forging an identity in a land new for their parents but not for them. To do this in Orange County, home of Disneyland, added a dimension all its own. Waves of kitsch flow outward from the park, so even a hotel that's a mile away has the soft, sloping roof of Minnie's cottage. And it's not like he didn't go. "If you grow up in Orange County, you spend a lot of time at Disney," Carlos says. Taking the "It's a Small World" ride was a set piece of cultural confusion. He didn't recognize his Vietnamese friends in the dolls moving amid gold temples. He certainly didn't see himself with a sombrero and colorful serape shoulder blanket. "When I took that ride," he remembers, "I identified with the animatronics."

By 1998 Carlos was working in the burgeoning Orange County computer industry, riding an early wave of programming, designing systems and writing code. That was a brief period in the tech industry when a single person could execute a project from beginning to end. "The knowledge base still hadn't grown so huge that there was a division between engineer, designer, and programmer," he says. "Soon every kid coming out of college had mastered Photoshop and Adobe Flash." He negotiated a good severance with his last employer, and when a group of his friends headed to San Francisco—gamers, most of whom he'd known for half a decade—he went with them. Not because he wanted to further his career. But he wanted to get out of Orange County. And he wanted a change. "My greatest accomplishments were virtual," he recalls.

The group settled into an apartment at Twenty-Third and Harrison. It was a typical gamers' pad, five people, twenty monitors. When his buddies went to work, he stayed behind, exploring the Mission neighborhood. He loved the deeply familiar smell when he opened the door of La Palma Tortilleria on Twenty-Fourth Street. He'd order tortillas, taken fresh from the rotary flat-top comal, and carry them back to the house. He liked the sight of the big, tropical papayas the street vendors sold by the entrance of the BART station and he'd take them back to the house also. Back, too, went skeins of fresh semolina and egg tagliarini he got from Lucca Ravioli on Valencia Street, which he might serve with ground walnuts and black pepper. Occasionally a whole fish from the Chinese fishmonger on Mission caught his eye. One thing was becoming clear: a bunch of hungry programmers were coming home in the evening and being greeted with very good dinners. (And Carlos was becoming less and less interested in programming.)

In the series of incremental steps that led to him becoming a chef, none may have been more important than his discovery of Harold McGee's *On Food and Cooking*: *The Science and Lore of the Kitchen*. Originally published in 1984, and eventually republished after a ten-year revision, the tome is the work of a writer adept in two worlds and able to bring them together in a single work. The first discipline dated to his time in Pasadena, where he studied astronomy at Caltech; the second one came from new Haven, where, in addition to a PhD in nineteenth-century English poetry from Yale, he developed a lifelong devotion to the thin-crust, coal-fired pies at Sally's Apizza on Wooster Street. The tome combines both an easily worn erudition with the certainty that scientific knowledge makes one a better cook.

Chefs were not the people who first bought the book. "They had their way of doing things," McGee recalled to me on the phone. "And they weren't about to change." Instead, the people who contacted him were culinary students. The letters still fill files in his home office. All wanted even just a little more knowledge or clarification on matters as diverse as the gelatinization of starch molecules or the role gluten plays in bread structure. Carlos's copy was soon underlined, cracked at the spine, and barely holding together. As an interest became a calling, he enrolled in the California Culinary Academy on Polk Street, gaining the confidence that led him to first apply to Winterland and now to Coi.

At the beginning, Carlos was one more cook in the cranking kitchen of the city's hottest restaurant. He wasn't sure if other than the initial greeting, Patterson even knew he was there. That was fine with Carlos—every young cook wants to keep his head down and prove himself—but it was also about to change. One day, Patterson was attempting to make an Arpège, a trembling barely set egg often served as an amuse-bouche, in a copper casserole on an induction burner. "He wanted to do it in a two-hundred-dollar Mauviel pot," Carlos recalls. "But the induction was not engaging with the copper." From his corner, Carlos mumbled something about how it could never happen, because copper isn't magnetic, and magnetism is how inductive heat is transferred. Patterson strained to listen, not quite understanding. Carlos grabbed a nearby fridge magnet and showed Patterson that though it was strong enough to hold up a sheaf of papers on the correct metallic surface, it would not stick to the pot he was using. From that day on, Patterson knew his name.

FINDING YOUR VOICE IS A curious thing. It's invariably more a winding than a linear process, so what might seem like an obvious influence doesn't throw an internal switch. After a satisfying combo platter at El Farolito on Twenty-Fourth and Mission, Carlos didn't decide he'd devote himself to crafting Mexican food. Instead of such a tidy trajectory, Carlos was about to enter that stage where disparate aspects of daily life lead to those things we want to articulate. Technical precision in cooking had given him an initial docking point, but the end result of his endeavor would be a new appreciation of the rougher, more immediate flavors tethered to his own heritage. As a child, when relatives visited from Mexico, they brought with them a mix of flavors that was foreign to him. They packed cheeses that were just a little tangier, concentrates of *mole* paste whose smell enveloped the house. And of course there were always candies for the kids. "These weren't candies like I'd known. They were vegetable or squash or cactus candies. I hated them then for their earthiness."

It might seem like at Coi he'd landed at the wrong restaurant to reconnect with that quality. Earthy? There were interludes at the restaurant that seemed downright effete! Patterson started meals with an aromatic component in the form of oils he'd devised with perfumer Mandy Aftel. Diners were supposed to place the proffered drop on their wrist—grapefruit was a favorite—and check in on its progress during the meal. That exercise may seem to exist in the higher reaches of sensibility—or pretension—but Carlos soon discovered if Patterson was doing anything, it came from complete conviction. He believed in the symbiotic relationship of the senses of smell and taste, hence the aromatherapy program. He believed in acidity, which he wound throughout the meal from the bracing opening sorbet to the reduced ribbon of homemade vinegar. "Whatever Daniel cooked," Carlos says, "spoke to his identity."

After a year, Carlos had become the pastry chef at Coi. He was now responsible for devising a suitable ending to the meal. These couldn't be what he calls "museum pieces at the end of eleven courses" but something that went with the rest of the experience—and its convictions. In San Francisco, great fruit was always an available ingredient. He'd internalized the slow rotation of the seasons, so he hauled through stone fruits and knew when the melons were coming in. He's still like that. "You're never ready for figs," he says with a laugh.

But the process of drawing from his background had started. Those were heady days and nights, cycling home after his shift, down the Montgomery Street hill and south on the streets of the Mission, his head full of recipes and possible combinations. Instead of plopping into his gaming chair, he'd now "get in late and try and pirate *Iron Chef* or Bourdain's documentary on Ferran Adrià," he recalls. A slow percolation was happening. Ingredients, memories of deep meaning, being reconceived as elements of fine dining. "There's a whole tradition of *ates* in Puebla," he says, giving an example. "It's a paste usually made of quince called *ate de membrillo*." The granular texture of the *pâte de fruit* he'd learned to make reminded him of a Mexican version made with guava. He had to call home to get the name. "I couldn't remember the word *guayabate*," he says, "a word I'd never bothered to learn."

In 2009, when James Syhabout asked him to be opening pastry chef at Commis, the Oakland restaurant he was planning, Carlos was surprised. "I didn't think I was good enough to be poached," he says, laughing at one of the rituals of the business. Syhabout had briefly been a chef de partie, responsible for one of the stations, at Coi but had left weeks before Carlos arrived. They'd met in the world of young chefs in San Francisco where the two bonded over their shared respect for Harold McGee's work—whenever the author ate at Coi he'd come into the kitchen for a quick hi—and soon realized they had much in common. They came from modest restaurant backgrounds. Syhabout's mother cooked in her small restaurant in Oakland, an immigrant-owned Thai version of Gregorio and Maria's La Siesta and, even if they left it unsaid, each of the young men knew what that meant. To be a child of restaurant workers is to grow up on a different tempo. You do homework on a Formica table or a crate in the back, your hour of play is snatched between services, you see exhaustion and resilience close up; you sense the good-night kiss grazing your cheek long after you're asleep.

That experience also conveys a very firm idea of the broader meaning of a restaurant. Humble though it may have been, your family's restaurant was a toehold in a new nation. Professionally, Syhabout's conception of that had started at David Kinch's Manresa in 2002 where he'd become a chef de partie soon after it opened. Kinch had already been running Sent Sovi in Saratoga for seven years when he first saw the ramshackle ranch house in Los Gatos that would become the celebrated restaurant. While working in Burgundy as a young cook he'd saved up to eat at some Michelin three-star places and was struck by "how each was unique to its location, like the five-house town of Mionnay, which I drove into and out of several times while trying to find Alain Chapel."

Josef Centeno, who'd come over from Kinch's Sent Sovi, was the opening chef; soon he was joined by Syhabout and Jeremy Fox, the Atlanta-born chef who would go on to redefine vegetarian cooking at Ubuntu in Napa. That was a development that would have been hard to foresee in the early days of Manresa. When not crafting exquisite amuse-bouches like green apple ice with olive oil or squid ink rice chips for customers to nibble on with a flute of champagne, Fox was developing the restaurant's charcuterie program. His particular fascination was the French blood sausage known as *boudin noir.* Crafting it required driving in his beat-up Toyota Camry to the 99 Ranch Market in Cupertino for milk jugs full of pig blood. "The first time I tried to make it, there was blood everywhere," Fox recalled to me. "Josef took one look and I think he walked out." The second time, using cooked rice to form the links, he served the sausage at the staff meal. After a few more tries, it was ready to be put on the menu, coasting in with roast suckling pig on a tiny mound of sauerkraut or, pureed into a robust sauce, spooned over a rabbit civet with sautéed foie gras.

The brigade working around the expensive Bonnet stove was driven and focused. Jason Marcus, today of Traif in Brooklyn, was there, too, and so for a while was Centeno's younger brother Aramis, fresh out of architecture school. It's worth being specific because the names of people who are together in a kitchen at one time tend to be how cooks remember distant days. It's a rapid business; daily prep leads to a rush of orders that cluster ominously during bad dinner services and become the distant past the moment the last dish is put out. New recipes are somehow snatched from the mania—scrawled out in pocket-sized notebooks with the intention of one day typing them up—other ones become ways of acknowledging mentors. Amid all the creativity, Kinch served a caviar-filled crepe drawn into a beggar's purse and tied with blanched chive. It was a salute to Barry Wine, who had created the dish at New York's Second Avenue temple, the Quilted Giraffe, and instilled core values in Kinch when he worked there in the '80s. After the night's service, when the Manresa cooks took sandpaper and polished the square stove until two A.M., it was out of a communal sense of métier.

That same sense radiated from Syhabout's restaurant when it opened. The very name, Commis, was the title of a lowly assistant just learning his craft. The lighting was recessive, the palette white and gray, the cooks worked in the pristine open kitchen wearing rough cotton aprons. Carlos was intent on justifying Syhabout's faith in him, and carefully calibrated meals at the restaurant often came to a perfect ending in creations like a signature black Mission fig tart, served with a perfectly shaped quenelle of lavender, almond, and beeswax-scented ice cream. He had reached a different level from the one he'd been on at Coi. He was no longer the young cook driven by aesthetics, but by relationships between ingredients and even between the restaurant and the community it existed in. And in Syhabout's vision, Carlos saw encouragement for his. "He'd call me out when it wasn't honest," he recalls of Syhabout's exertions to always consider how the small restaurant was interacting and reflecting with its broader environment. That drove Carlos to forage for wild fennel, borage, and lemon balm for sorbets in the hills above Oakland, and it made him respect the $68 price for a five-course tasting menu that reflected Syhabout's wish to have fine dining remain accessible to those who might not otherwise try it. Reviewers took note of this effort and perspective. "Any restaurant can fill you up," Josh Sens wrote in *San Francisco* magazine. "Only a rare few leave you this fulfilled."

I PUSH MY STOOL BACK from the counter at Taco María. Carlos is busy; his back is to the dining room as he shows one of his cooks how he wants something done. I've come to know this mall. During the day, there are olive oil tastings at some stores, coffee roasting at others. Taco María brings people in, too, and at weekend brunch the counter and terrace fill with towheaded families eating his layered parfait and *molletes,* rafts of grilled country bread freighted with a sauce of blistered tomatoes and chiles. At night it gets quiet though, very quiet. I walk out to the large parking lot and start for the freeway. Each time I approach the 405 on-ramp I'm struck by the knowledge that if I were to head south instead of north, I'd be at the Tijuana border crossing in less than two hours.

That geographical proximity played an indirect role in Carlos's return. It underscored a tension that, hesitantly, he found himself wanting to address. He wanted his restaurant to be rooted—but what were his roots? And how would he frame what he was trying to express? It was one thing to charge seventy dollars for a tasting menu in Oakland, but in the restaurant he'd grown up in, you gave away big baskets of chips, and some customers got ticked off if you didn't give them a free tub of salsa to take home as well. But Carlos came to understand that having such a heightened awareness of perceptions made it all the more incumbent on him to challenge them. After discussing it with Syhabout, he publicly announced he'd be leaving Commis in May 2011. Staff changes in Michelin-starred restaurants get immediate coverage, and the *Chronicle* wanted to know what he'd be doing back in Southern California. Salgado had an idea for a restaurant, the paper reported, "which will specialize in what he calls 'first generation Mexican-American food.'" That "what he calls" is rich. It really means, *we have no idea what he's talking about.* Carlos wasn't all that sure either, yet. But he knew he'd found the expressive place inside himself. He could return to Orange County with his knife roll. He was a chef who was ready to be heard.

CODA

I see the word *artisan* everywhere. It jumps out at me from paper coffee cups at my bagel spot and the back label of a wine bottle on the table. I hear it touted on TV ads by fast-food chains eager to take on some association of tradition. I used to sneer at the national pizza joint I'd see getting frozen goods wheeled in from an eighteen-wheeler while touting its "artisan" pies, but that kind of abuse of the word doesn't bother me anymore. I'd rather see it out there, used—and even at times misused—because of its evocative power than relegated to an archaic term.

The word has done its job, allowing us to talk about the different ideas that hover around flavor, such as integrity and intent, otherwise hard to get a handle on. And it's not something that can be announced, anyway. You never really know when you'll feel the power of an artisanal moment. Perhaps one random morning you notice the patience with which the pour-over coffee is prepared at your local spot. Or perhaps it's while taking an open house tour of a producer, when the guide has already moved the group along and you delay for a second in the small business park hangar realizing the casks around you are transforming a cider of local apples into deliciously brisk vinegar. Rhythm slows down and we're in that space where nothing has to be done. The process has to take its course. It can be powerful. This is a value from another time somehow completely vital in the world of today.

Thoreau had a word for that. In *A Week on the Concord and Merrimack Rivers,* the journal of a boating trip he took with his brother, John, in 1839, he described the way ancient forests jutted into the newly growing towns as "inspiriting." The world was changing all around—the canal would bypass the river, just as the distantly heard trains would eventually bypass the canal—but here in these old-growth trees was a reminder of permanent values, perhaps even of who we were.

If you add a little whimsy, you're getting close to what real artisanship means. It begins after all as a craft, one that has to be learned, often the hard way. "We lost twenty cases learning to *degorger,*" Tracey Brandt of Berkeley's Donkey & Goat Winery once told me of their early efforts to make bubbly wines in the most traditional way. That's the kind of hell-bent enthusiasm that excites me. Everything is learnable if you have the right approach, some modesty, and perhaps a mop. Eventually, people who've subjected themselves to this discipline turn those hard-won skills into an animating force that invigorates our daily life and even American cities.

In the course of writing this book, I have amassed files that bulge with clippings of ventures I felt I needed to get to. I have the names of brewpubs in Columbus, coffee roasters in Cedar Rapids. There's a farm-to-table restaurant in Omaha. Zürsun heirloom beans in Twin Falls, Idaho, Crooked Stave Artisan Beer Project in Denver—how could I have written this book and not been to any of these places yet? Of course, the widespread abundance of places driven by excellence actually represents the thoroughness with which the country has been transformed by an artisanal approach. It's a matter of something powerful being compressed into a particular flavor, something great—and, yes, inspiriting—reached through a delicious bite or sip. We don't have to fly anywhere to tap into it. It's there in the flight of single-source chocolates we share with friends, as we watch a cheesemonger cut into a large, aged wheel, in the sound of good bread crackling as we break it with our hands.

ARI IS SITTING AT A trestle table outside Zingerman's Bakehouse in the outskirts of Ann Arbor. He's got a laptop in front of him and he's intermittently sending e-mails, perhaps even working on the next newsletter. Wearing his customary black T-shirt, black jeans, and black Doc Martens, he's immediately recognizable. Two business students who've bought his book, *A Lapsed Anarchist's Approach to Building a Great Business,* have wandered over with hearth-baked bagels they've bought inside and he signs their copies for them. "Lead well," he writes, and he asks them about their courses.

Zingerman's *is* a great business. At the deli, still at its original downtown location, there's probably already a lunchtime line creeping out the door as people wait to put in orders for sandwiches with cute names like Tomorsky's Temptation, a mustard-slathered Kaiser roll stacked with kosher-style salami and Wisconsin Swiss. Another Ann Arbor venture, Zingerman's Roadhouse, sits surrounded by a large parking lot. It looks almost like a chain restaurant, except coffee is served from a repurposed Airstream trailer by the entrance, the corn dogs are dipped in

Anson Mills cornmeal, and the vegetables are from their own farm. Open all day, it lends itself as a meeting place for the community. As much a greeter as he was in the old days at Maude's, Paul sometimes does a goodwill tour of the tables in a porkpie hat, filling the rooms with an easy laugh. Ari shows up, too, doubling as a busboy as he splashes ice water into glasses from a pitcher and exchanging a few words with regulars.

Out here by Ann Arbor's airport, things are quieter, particularly now on a late June day. This small office park on Ann Arbor's southside is a fairly anonymous setting for the cluster of businesses the deli has created. There's the coffee roaster with a cupping room and tables that look into the area where the freshly roasted beans can be seen spinning in a bright red metal cooling drum. Farther down the concrete slab of pavement (it makes for easy scraping in winter) the small creamery sells imported and American cheeses as well as its own cream cheese, inspired by how cream cheese was made in small dairies in the past. NO VEGETABLE GUM, HAND LADLED AND DRAINED FOR SIX HOURS reads the sign above the tubs with typical thoroughness.

That's the kind of detail customers have come to expect from Zingerman's. From the beginning, Ari and Paul knew that people had no reason to buy little-known Irish tea bags, Tunisian olive oil, vials of hundred-year-old balsamic vinegar, or even Nueske's bacon. They had to give them a reason to do it. Throughout Zingerman's literature, information is presented in such a way that it doesn't talk down to those who might know it, yet it informs those who don't. The company's tone—the Ari tone, really—is full of fact, on your side and always colorful, inviting you to join him down some country road, peering into some dark farmhouse where a powerful aroma hangs in the air "as the olives are crushed, the green-gold oil pouring out of the separator (the final stage of the process) onto a slice of thick country bread."

The "look" is just as carefully managed. In a low-slung building directly behind the Bakehouse, the art department is working on the next mail-order catalog. When they started to figure out sales beyond the store in 1993, they offered local deliveries with a modest charge, ten dollars for Ann Arbor, an extra five for delivering to nearby Ypsilanti. Now they deliver from coast to coast, an enterprise based on a catalog whose every page is an enticement. The drawings have an intentional wonky, slightly jostled look. Perspective, after all, is not important; approachability is. It takes a lot of skill and hours to look so casual (and edible). Though the four people in the art department work standing at their monitors, art director Ian Nagy sits at his desk. Carefully mixing colors with the tip of a paintbrush, he paints a slice of a special cake, Hungarian Cardinal Slice, on heavy paper board. The way he traces the checkerboard of crisp meringue and sponge cake containing clouds of whipped cream boosted by a fillip of crème fraîche conjures up all the off-sweet wonder of classic Hungarian baking. Of course you want a slice.

Much has changed in the three and a half decades Zingerman's has been in business. At the deli, the sandwich orders are taken on an iPad, and the Muno Bold font comes out of a laser printer. They're no longer interlopers in the halls of gastronomy, they're leaders. "When Ari goes to the Fancy Food Show now," says Paul, "he gets treated like a rock star." We're sitting over pancakes early one morning at the Roadhouse and Paul is in a talkative mood, putting the decades into context as he takes sips of coffee and intermittently scans the room. "No one was doing menus on chalkboards when we started, no one gave samples. The owners didn't even give the employees samples. If they caught you eating something, they called it stealing." Paul is right; deli retail has been transformed by the Zingerman's way of doing things. We expect to be invited to sample just as our enthusiasm is boosted by that of the staff. The term *fancy foods* even sounds dated because it's the way that we eat. The Zingerman's crew was always goosing the pretensions of the term with tongue-in-cheek promotions such as using the occasion of Prince Philip's sixty-fourth birthday or the Yodeling Festival at Interlaken to offer 10 percent off all English or Swiss cheese.

But something serious has also gone on, something that is just as representative of what the food world has become: they've found their place in the world through it. That has meant Zingerman's consciously deciding to not expand outside Ann Arbor. They never saw it any other way. In 1988, Ari told *Fancy Food* magazine they didn't want "the additional headaches of managing multiple locations." Today, when you stand in the deli, look through the well-stocked shelves, or are offered a taste of something, it feels like a unique place. Paul is aware of how increasingly special that sensation is. "When you go to the original Starbucks in Seattle, you really feel something," he says, taking a swig of Zingerman's dark brew. "When you go to the Starbucks on the Ohio Turnpike, not so much."

For Ari it has meant the opportunity to keep on learning and to deepen his understanding of what good service means. Training was always important. Even the first employees were given a three-ring binder. That grew into the booklet employees receive today, one that is direct and empowering. "Do not argue the point" is the first of four steps for handling customer complaints. The final one grants permission to "do whatever you think is necessary to make a dissatisfied guest happy." Ari has developed that interest in providing the best service into hugely popular seminars on staff training and visioning, a business model that leads people through the steps of articulating their goal first and then setting up the steps to get there.

For all his teaching and writing, Ari is likely to offer a personal insight in terms of food. "Growing up in a kosher home, bacon was way out of bounds," he writes in his book *Zingerman's Guide to Bacon*. "While we ate meat in restaurants, pork was always completely *not* OK. Warnings and admonitions were unnecessary: not eating pork was as given as a given could be." That's a nice introduction to discussing the light smoke practiced at Nodine's in Goshen, Connecticut, the nitrate-free style of curing Niman Ranch practices, or Bruce Aidells's opinion of the "lean-to-fat ratio" of the Duroc and Berkshire breeds. It's also typically warm and atypically personal. Ari goes even further into what a life in food has meant to him when he salutes the writing of a little-known British bacon curer he admires. "One of the beautiful things about the *Bacon Curer* books is that Maynard so beautifully captures the essence of finding a vocation. Unlike the more common approach to work ('TGIF' and all that), a vocation is when we work at something that makes us feel truly fulfilled, and which we go at with great passion every day. And we do it, not for short-term material benefit, but because we feel positively wedded to the work, at peace with the processes at play and fulfilled for following them through to fruition. It's honestly the way I feel about my own work."

LA BREA BAKERY *HAS* EXPANDED. Today it is owned by Aryzta, a Swiss company, but Nancy Silverton remains attached—the story of how she started the enterprise in 1989 is still printed on every one of its maroon bags in all sorts of packaging formats, including economy-sized ones for dinner rolls at Costco. The company's delivery trucks are a fixture in the early-morning landscape of Los Angeles. Purring with an electric motor, the trucks pull up outside cafés, restaurants, high-end burger places. The sight of the drivers ferrying the orders in the dawn light captures the old bakery truck delivery but for a modern age.

There is a value to that expansion and there is a price. Some people hold ubiquity against the company. Too big, they say. And there are times when the interior of one of the supermarket baguettes (parbaked and finished in the market's ovens) doesn't have the full, gorgeous crumb of La Brea's former bread; still, its broad reach has raised the bar for everyday bread enormously. I use the parbaked dinner rolls to make the sandwiches my children take to school. There are days when my first coherent thought is mumbling the well-memorized instructions on the back of the package—"385 degrees for five to seven minutes"—as I press buttons on our oven.

One morning, I make an appointment with Nancy to go see where it all is made, a large plant in Van Nuys. We meet at the new retail bakery, a block up La Brea Avenue from the original location. They still use the hand-lettered names for each bread her father, an amateur calligrapher, drew. She points this out to me, seeming pleased. There are people ordering coffee and pastries, and she feels like a visitor herself. We hop into her black Porsche Carrera and drive out toward the Valley. She negotiates traffic like a native. Her life is different now. A partner in the high-profile Italian restaurant Mozza with Mario Batali and Joe Bastianich, her fascination with the single ingredient has transferred to fresh Italian cheese. Her work ethic is strong as ever, and most nights she can be found at Mozza's marble-topped horseshoe-shaped cheese counter, dusting ricotta with lemon zest, dishing up flown-in Puglian burrata.

When I called and asked if I could see the big plant, there hadn't been a moment's hesitation. No conferring about what it might mean to let a journalist in, no slanting of the story to when the business was tiny and hence more promotable. Instead she'd said, "Let's meet at the bakery around ten. Traffic won't be too bad then." Her hair is up, her sunglasses on. I admire her directness. She once took the racks out into the parking lot of the bakery to cool, but it would be bogus to pretend she's doing that now, and she doesn't. The plant is out near the Van Nuys airport, where train tracks cross under the 405. We've barely been buzzed in and donned our hair-nets when I get to see what large-scale baking means. Flour is kept in silos, bubbling yeast slurry in bins, rolls pop out of an oven on a conveyor belt, tumbling into cooling troughs. The sound, cacophonous and constant, is the sound of industry.

Nancy walks the floor with me, listening to the head technician, pressing a knowing finger to a batch rising in what looks like a mine cart. Whenever she's spotted by someone she knows, they give her a warm hug. They're mainly Latino men and they've been baking now for a couple of decades. They catch up. "El Tigre?" "Went back to El Salvador." She mentions someone else. That one retired. "Retired!" Nancy says with a laugh. This is the side of scale that no one sees, and in a way, more than any silo full of flour it's what I've come to observe. The hum of the conveyor belts and ovens is constant, but to me that noise has a positive undertone, an economic one, of homes being bought and kids being put through school. Nancy is pleased as we leave and head toward the freeway. I can tell by the ease with which she drives.

IN 1985, ALYCE AND DOUG left Michigan for Alabama. Farmstead cheesemaking, the kind of operation they wanted to start, had little meaning in the agricultural offices Doug was obliged to take triplicate forms to. He chafed; this was his home state. But in Alabama, there'd been a lot less red tape, and the weather made it a far more attractive place to live. In the eyes of the authorities, here were people who wanted to farm; they might not be the kind of folks that usually came around the Ag Bureau, but with farms shuttering all around, it was enough that they wanted to put cows out on land. In all, it took three trips. The first was exploratory; on the second they bought a spread from one of those farmers who had been hanging on. They knew they'd found the right place the moment they saw the open field with its stand of pecan trees that would offer some shade to the Guernseys. On the third trip they loaded the cows in a trailer and drove forty-eight hours. When Alyce made her first batch of cheese, it was in a landscape that even bulk milk hadn't been able to save. Sweet Home Farm was the first farmstead cheesemaker in Alabama and the only working dairy in Baldwin County.

The town of Elberta is located east of Mobile, in that narrow strip between Biloxi, Mississippi, and Pensacola, Florida, that represents the Alabama coastline. The white sands of the beaches attract vacationers to towns like Gulf Shores. Elberta is inland amid country roads shaded by pines. A hand-painted sign reads NATURAL FARMSTEAD CHEESE, and an arrow underneath points visitors toward an unpaved path that runs half a mile to the tractor shed, house, and cheesemaking room. Alyce is doing chores when I arrive. She only sells on Friday and Saturday, and because of spring break, the next day she's expecting a rush. "Our locals know exactly what they want," she says, and cuts me a slice of Perdido, a firm cheese she makes layered with ash from homegrown herbs. That gray ribbon running through the center tempers the cheese's lushness. "Do you sell a lot of this?" I ask.

"That one's real popular," she answers firmly, just in case I'm insinuating something about Alabama.

An activist from way back, she's very involved in resisting any further government regimentation of cheese. Her cheese is not sold before it has aged for sixty days, so it doesn't need to be pasteurized. But there's talk that even that might be extended by law. "If I had to pasteurize the milk," she says, "I'd stop making cheese." We step into a shed with chairs and a table to talk, and soon Doug arrives with slices of cheesecake Alyce has made from the milk. It's incredible; this time the well-baked crust performs the task the ash does in the cheese: parry the richness. With her hair up and reading glasses hanging on a cord around her neck, there's something poised about her. But Doug soon has her laughing. "When we first came down here, we used old five-gallon buckets from Piggly Wiggly to mold the cheese," he says, and she shakes her head at the recollection.

She shows me around the property before I take my leave. The cows are out among the pecan trees, the garden is growing, there's plenty of old farm equipment around for Doug to keep busy with. I get a real sense of permanence, even though the couple started as uncertain homesteaders in the small town of Fennville, Michigan. It wasn't a lark; it was a life.

MICHAEL DEMERS HAD NEVER QUITE experienced cooking the way Jean-Louis Palladin envisioned it. A chestnut soup began with raw chestnuts that had to be scored, roasted, and peeled. And they weren't even the main flavor. "Then you got the mirepoix to sweat them in," he tells me on the phone. "Prosciutto ends and foie gras." Demers had been Jean-Louis's chef at Napa, the restaurant he operated in the Rio Hotel and Casino in Las Vegas from 1997 to 2001. The Watergate restaurant was over for him by then but Jean-Louis didn't look back. He enjoyed the desert city, the rush of ending the day at three A.M. eating wok-fried Dungeness crab alongside dancers from the shows on the Strip. At the quieter places, like Rosemary's on West Sahara Avenue, he drank port happily at a table by the door.

He'd been brought to Las Vegas to launch a fine dining restaurant as part of the Rio's makeover. The casino's owner, Tony Marnell, was determined to make the property a destination, constructing the two-story Masquerade Village, a retail-entertainment expansion with an array of dining options. Napa, Jean-Louis's restaurant, was on the second floor above the seafood buffet. He'd convinced Marnell he needed a few simple things. One was a wood-burning oven that could crank to a thousand degrees and with it of course a constant supply of olive wood. He didn't care where it was stacked, but operations had to know he wanted his cooks to have access to it. They agreed it would be stacked in a sheltered section of the parking lot. "He had a way of cooking the steak that was fantastic," recalls Demers. "He'd heap coals in a pan and put a grill over that and the steak went on top and the whole thing went back into the embers. We did whole foie gras like that, too, with shallots and thyme."

He is warmed by the memories, but I know the moment will come when there's a catch in his voice. There always is when talking about Jean-Louis. He was diagnosed with cancer in 2000, he died in November 2001. A short time. Michel Richard remembers the last days when Jean-Louis couldn't move for the pain.

Jimmy Sneed, his right hand in D.C., puts it with a hurt anger. "I'd toss out his damn cigarettes." Daniel Boulud, whom Jean-Louis took under his wing in 1982 when Boulud had just arrived in the United States to be private chef at the European Commission, allows himself a long pause when I ask when was the last time he saw his friend. He mentions something about Sloan Kettering, then trails off. The toughest voices to hear the catch in are the former apprentices. Speaking from the One & Only Palmilla Resort in Los Cabos, Mexico, where he is chef, Larbi Dahrouch wants to keep to good memories: Jean-Louis's face inches from a plate as he applies the finishing touches. Running kitchens for Alain Ducasse would be among the many achievements of Sylvain Portay's career, but he was fifteen years old when his parents signed a two-year contract and he joined Larbi and Jean-François Taquet in the apprentice corps at La Table des Cordeliers. "We'd be playing pinball at the Café des Sports and he'd come in and yell at us to get to work," he remembers when I call, as if still hearing the gravelly good-humored voice. I ask when he knew Jean-Louis was sick. "It was quickly known," he says. "We did a benefit for him at the Ritz-Carlton in San Francisco. All the chefs were there. Michael Mina, Laurent Manrique, Chez Panisse." He's speaking quickly as if to outrun the catch. Then it comes.

But it's important to remember Jean-Louis in another way. The connecting link to all the anecdotes of him tracking down farmers and fishermen and people who raised squab or grew chervil just for him is that he changed the game. He transformed the role of chefs from passively waiting for delivery to actively searching for what their cooking required. He teased out the tension between haute and rustic that always existed in French dining and gave it an American shading with American goods. There was Maine seaweed in the salads; compound butters made of lobster coral ran down the seared sides of Nantucket scallops; shad roe got wrapped in caul fat before searing and fanned out on fine china before being drizzled with bacon vinaigrette. This was still the era in this country of dining rooms rotating above corporate towers. The soup that started with sweating off the kind of muscular mirepoix he liked to use as foundation, adding, say, mushrooms foraged for him in Virginia forests as a second step, ended as a seamless blend of dining's high-low registers because Jean-Louis never used plain chicken stock—instead, he devoted the time to making clear consommé. No one would ever see its hard-earned clarity once it got utilized as a stock, but that wasn't the point. "It gives the food a beautiful flavor," he told an interviewer.

I never got to taste his food, even though I live and orbit in a restaurant world of robust flavors and demanding sourcing that he brought about. After all my reading and interviews, I wanted to experience something living about his style. I asked Josef Centeno to make a batter he'd once told me came from Jean-Louis. I went to downtown L.A. and watched him prepare it in his American-inspired restaurant, Ledlow. "I got it from a friend who got it while working for Guenter Seeger in Atlanta who got it from his friend Jean-Louis." It was all simplicity, a combination of flour, baking soda, cold bubbly mineral water. Josef chose to fry zucchini flowers stuffed with homemade pimento cheese. Once he'd filled the flowers with a pastry bag, he dipped them in the batter but he didn't drop them into the hot oil; instead, he lowered them with a figure-eight movement so they'd form a cap and keep their shape. Eventually, he served them drizzled with sorghum syrup. They were delicious. But it was the moment of seeing him skillfully lower the flower into the oil instead of pitching it in that reminded me of the intelligence of a cook's hand. I thought Jean-Louis would have appreciated that moment. I almost felt him there right then.

PRECRUSH, THAT PERIOD WHEN THE grapes haven't yet been picked, and I'm getting to know Carlton, Oregon, well. The tiny town in Yamhill County was originally named for a Mr. Carl who in 1875 convinced the Westside Railroad to establish a stop where he'd settled and Westside named it Carlton after him. I got that nugget from the copy of *Reflections of Carlton* published by the town's elementary school Bicentennial Club in 1976 that's in the library collection in McMinnville, a nearby, slightly bigger town. I spend my share of time there. Also at the Subway franchise. When not doing that, I wander the streets of Carlton itself. I have gazed at the one traffic light that hangs suspended at the main intersection right near the Madsen grain elevator. It's just a few steps from the little community pool, drained now in early October. I guess I'm avoiding lurking at Ken Wright Cellars, a gray clapboard building fronted by a trellis in the center of town because my presence reminds Ken he's waiting. Ken flits in and out of there, already unshaved, already wearing the hobnailed boots he'll barely take off once the picking starts. But it hasn't started yet. The sorting line the grapes will be inspected on is clean and checked, the fermenting vats are empty and marking time. Everything that can be prepared has been. He could pick but he's not ready to. If he has to see me prowling around, one more reminder that an entire mechanism is waiting for his word, it might make him flip.

He's playing a waiting game. He's gambling he can gain more nuance from the "hang time" the grapes spend on the vine in the face of the fall storms that can unfurl with the speed of a bullwhip cutting through British Columbia and through the Puget Sound and delivering a last destroying flick to Oregon. To his harvest. "If your fruit isn't in the barn by October fifteenth, you're going to get hit," he says. It's October fifth.

From time to time we get into the dust-covered Ford F-350 "Stump-Puller" he drives and go up into one of the nine pinot noir vineyards he picks. Some, like Canary Hill, are very pretty, nestled amid growth of fir, oak, and manzanita. The order he picks the vineyards in will be determined by the maturity of the fruit, which invariably is a reflection of the soil they're planted on. The soil of the Guadalupe vineyard is Willakenzie, a well-drained shallow soil over a base of siltstone. "Guadalupe is always first," he says, "even among the sedimentary soils."

There's a refractometer tossed in with random papers in the truck's cabin, but I've yet to see him use it to burst a grape and get an immediate sugar reading from the instrument. Instead he walks between the trellised vines, perhaps plucks a leaf, perhaps talks with the owner if it's one of the vineyards he has a contract with. He knows what he wants. After all the soils and all the rootstock clones have done their thing, after the trellis system has protected the clusters in summer's heat and translated every last half-tone of fall light into fruity potential, each of the vineyards will communicate place. That is his entire approach to wine. That means in October, you pace. On Canary Hill (a mix of Jory and Nekia soils) he takes a grape, tastes it, and spits out the skin. "We're running out of heat," he says, looking at the sky. The loud report made by a carbide cannon firing on a timer to keep birds away is the only other sound.

Of the nine vineyards he works with, the one I know best is

Abbot Claim, which is not far from town, down a narrow road lined now with tall drying grasses. I've known that vineyard for almost a decade, as long as I've known Wright. During an early trip, we'd driven up to the southerly facing hill bordered by blackberries and ponderosas. There were no vines then; a backhoe had dug a crater at the center. Ken stepped over the yellow caution tape, hopped down, and ran his hands along the exposed cross section of earth. Certain layers were streaked with gray, others were reddish clay. Others held tiny pebbles. "You want the vine to get to the mother rock," Ken had shouted up excitedly, "because that's where the minerals are. That's the point when the vineyard starts showing clarity of character."

A few years later I'd made a second trip to attend the International Pinot Noir Celebration held on the leafy grounds of a college in McMinnville. We'd peeled away during an afternoon lull and gone to visit Grant Youngberg, the elderly man who lived in a farm on the valley floor right below the Abbot Claim vineyard. Well into his eighties by then, Mr. Youngberg had talked about how his family had received their land claim, in an earlier version of the process that had granted John F. Abbot his 160-acre homestead. I'd seen the map put out by the Surveyor's General's Office in 1867. If it wasn't for William Smith's claim directly to his north, Abbot would have been surrounded on all sides by men with the kind of old biblical names that transport us to another era. Abijah Hendricks's claim extended to his north, Eli Perkins homesteaded directly to Abbot's south, Ruel Olds was one claim away, and Absalom Hembree's property bordered Abbot's on the west. With a glimmer in his eye, Mr. Youngberg told us to go out back, open the paling gate, and walk into the waist-high oats. We did. Absalom Hembree was buried there. Yamhill County's first sheriff.

Ken had been excited to hear him talk as we returned to the kitchen. Mr. Youngberg wasn't talking wine but, just maybe, something that went into it. He was playful. "I have to be careful of what I say," he'd said, laughing. "I'm related to half the county." He could remember the days of Carlton when the sawmill provided the electricity. "At 10 P.M. they turned off the electricity," he explained. "At 9:50 they blew a whistle to warn people." But he could go further back, returning to the settling of the land. At one stage he went upstairs and came down with a low, four-legged chair with what looked like straps of well-worn hide providing a latticed sitting area. "This chair came across the Plains in '44," he said. Meaning on a wagon train in 1844. It was small enough that he put it on the Formica table and we all gazed at it, a special object amid the vitamin supplements and the *Reader's Digest* copy that composed Mr. Youngberg's everyday.

I understood what Ken was trying to do. This was inhabited land. This was not the blank slate that people talk about American wine coming from. It once had belonged to the Yamhela people, and they lost it by force, and then a party of surveyors had come out from Oregon City hauling the Gunter's chain with which, in units of ten square chains to the acre, it had divided the territory into meridians, ranges, townships, and land claims. The hillside slope Abbot had been deeded was not excellent for crop farming—at one stage in the 1970s it had been a Christmas tree farm—but its exposure and soil made it ideal for pinot noir. I walked out of there as excited as Ken did. Here was the sense of place that wine tries to capture.

What I respond to in Ken is he doesn't come at the idea directly in conversation. *Terroir*—it's the concept behind everything he does, but he sees it in its broadest way, not a hillside but a living, thriving place such as Carlton. The town where the sawmill had rung the warning for the end of the day is holding its own against franchise sprawl while still remaining viable. Ken Wright Cellars is a big part of it, helping sponsor—along with the body shop, the farm store, the local Hallmark stationer, and other businesses—the annual Fathers and Daughters Ball. It's held at the winery, and there's plenty of pink bunting put out, and some serious nail painting and glitter dusting is undertaken at tables amid the stacked barrels.

But now that room is clean, spotless, empty, and waiting for his word. Isolated like this, flavor becomes an abstraction. The unseen Godot character that drives the story. I've had to get over the fact that because of Ken's dedication to flavor I won't get to see the harvest, the purpose of my long-planned four-day trip. As I pack my bag, flavor seems like the rude guest whose tardiness upsets everyone's plans. But I'll get over it. Each time I come up I scratch at a layer of the area's individuality. Before I leave I stop in at the winery. Ken is upstairs in the office looking—gazing—at a massive At-A-Glance Universal wall calendar covered in vineyard names and numbers written and rewritten in Sharpie. His baseball cap is off and his hair stands straight up, one of his winning features. He smiles ruefully, gives me a slap on the back. "Next year you're coming up to work sorting line," he says.

"Say no," one of the office workers says. "That's real work." As I drive out of town I think I detect just a slight hint of heat trapped in the breeze. Definitely worth waiting another day.

IT'S A MUGGY SUMMER DAY when I visit Orb Weaver, Marjorie and Marian's farm north of Middlebury, Vermont. They've told me to look for a mile marker a few miles before the road jogs west and continues up past Vergennes, running alongside Lake Champlain. The road they share with other farms is narrow but gorgeous, each curve giving a better perspective of Mount Abe, Camel's Hump, and other landmarks in the foothills of the Green Mountains.

Marjorie and Marian are out working in their vegetable garden. They're wearing well-worn shorts and T-shirts, and they finish up tying tendrils of tomato plants to bamboo poles to keep them climbing before showing me around. They've made a few concessions. They stop their cheesemaking a few months earlier than they used to every year, and now, when a calf is born in the farthest parts of their pasture, they'll put it in a wheelbarrow instead of carrying it all the way back to the barn. Marjorie had a hip replacement surgery recently. "That's what you get from schlepping around sixty-pound bales of hay," she says.

But the farm is how they make their living. They still fill their old Toyota with the lettuce, peppers, eggplant, and shallots they grow and drive to the Middlebury co-op as they've been doing twice a week since 1981. One of their specialties is a basket of different-colored cherry tomatoes, and they spend a lot of time sorting them so each pint basket has a nice mix of hues. When they make cheese, they work as one, knowing each other's rhythms, tamping the curds down, weighing, setting them into the press. They deliver the ten-pound wheels to places like the Healthy Living Market in South Burlington, where it's put out with Shelburne Farms cheddar and maple-smoked Gouda from Taylor Farm, all within an hour's drive. Occasionally someone will stop them and say, "We were brought up on your cheese and now we're bringing our kids up on it."

That permanence was formalized recently when they signed

over their 103-acre farm to the Vermont Land Trust. It will always be farmland. "It's a good feeling," Marian says. "People come up to us at the farmers' market and they thank us for doing it." They used the money they received to fix the barn. It's an old nineteenth-century structure with a high, sloped roof but it was really on its last legs. They stabilized the entire foundation so it could stand for a hundred years more.

From this vantage point, the artisanal revolution is a movement that's allowed people to remain on the land and make a living. Because of people like Marjorie and Marian, the American landscape has remained a working landscape, not just a pretty one. We've returned to the garden and sit out in chairs among tall, blousy flowers, not noticing that the sky has darkened. At least I haven't. Marian looks up with an experienced eye; could the storm do any damage? "We get those," she says as a bank of gray clouds funnels toward us down the Champlain Valley. It's a real Vermont storm; it skips the drizzle stage. We're suddenly moving through the rain, laughing at the speed with which the downpour came. "Hurry," Marjorie calls. I run behind her. She's wearing old high-tops and a tie-dyed shirt. The wind is lifting her silver ponytails like a young schoolgirl's. Moving slower, Marian is also amused as we run toward where Marjorie is holding the screen door of their home. "We'll wait it out inside," she says. "Marjorie made bread."

ACKNOWLEDGMENTS

When Kit Rachlis asked me to be restaurant critic of *Los Angeles* magazine in 2000 he gave me a front-row seat to a transformative time in American dining. Social media and the cell phone were about to jostle expectations of a restaurant meal and recast the role of diners, even as chefs broadened the possibilities of what cooking could express. Observing this national trend from the vantage point of Los Angeles—a city that from the wood-fired open kitchen of Spago in 1982 through the Kogi truck lines twenty-five years later has always been driven by the new—continues to be a privilege. It was Kit who first sensed there was a greater story to the resonance ingredients had taken on and I am grateful for the enthusiasm with which he communicated that such a project was not only reasonable to consider but possible to accomplish.

My monthly columns have benefited enormously from the deft editorial touch of Matt Segal, who additionally helped untangle a particularly jumbled section of this book. His key question, "But what does it taste like?" hovers over my mind pretty much daily. Editor in chief Mary Melton has been warmly supportive of my dual work reviewing and researching. I am grateful to dine editor Lesley Bargar Suter for the enthusiasm she had for the project throughout. I also thank colleagues Daniela Galarza, Linda Burum, Randy Clemens, Bill Esparza, Garrett Snyder, and Josh Scherer. In addition, Chris Nichols has been generous with his knowledge of the San Fernando Valley's history, while Emily Young read the manuscript with an attentive eye for mistakes, though any that remain are on me.

It has been a special privilege to conduct research at the Huntington Library in San Marino. I am grateful to Alan Jutzi for believing in the book (and granting me a reader's card) and guiding me to both the Anne Cranston American Regional and Charitable Cookbook Collection and the Robert V. Hine Collection on Communes and Utopias. In addition, Dan Lewis pointed me to the library's collection of Pasteur papers. I am indebted to Leslie Jobsky, Frank Osen, Michael Fish, Alisa Monheim, Jaeda Snow, Maria Blumberg, and Sarkis Badalyan. The feeling of collegiality extended throughout my research at the Huntington and I am particularly grateful for the friendship and encouragement of fellow researchers Helene Demeestere, Barry Menikoff, and Richard Mandell. The Los Angeles Public Library has played an equally important role in my reading and I have benefited from the goodwill and patience of those librarians who oversee the culinary department, in particular Jack Stephens.

Many individuals spoke to me repeatedly and patiently in person and on the phone, generously sharing their memories or the details of their craft. I am greatly indebted to Nancy Silverton for recollecting her early days in baking. Alyce Birchenough, Doug Wolbert, Marjorie Susman, Marian Pollack, Ricki Carroll, Judy

Schad, Andy Hatch, Scott Mericka, Mike Gingrich, John Jaeggi, and Gary Grossen have been particularly helpful on the subject of cheese. Axel Borg and John Skarspad of the Peter J. Shields Library at the University of California, Davis, directed me to the papers of Maynard Amerine. Also generous with their memories of Dr. Amerine were Harold Olmo, Ralph Kunkee, Warren Taylor, Robert Balzer, and Darrell Corti. Dan Linscheid gave me an overview of the history of Oregon surveying. Ken Wright, Rollin Soles, Allen Holstein, David Adelsheim, and Josh Jensen were generous with recollections of their early days as winemakers, as was the late David Lett. I am grateful to Frank Carollo, Amy Emberling, Steve Wallag-Muno, Pete Sickman-Garner, Ari Weinzweig, Paul Saginaw, Dennis Serras, Ian Nagy, Eric Olsen, Allen Leibowitz, Jan Longone, and Maggie Bayless for their memories of Zingerman's beginnings.

I would especially like to thank Régine Palladin for talking to me about Jean-Louis Palladin and to Phyllis Richman for recounting his early days in Washington, D.C. Also generous with their time were Ariane Daguin and her father, André, Larbi Dahrouch, Sylvain Portay, Bill and Joan Burgess, Roy Young, Roger Yaseen, Jimmy Sneed, Rod Mitchell, Nicholas Branchina, George Faison, Daniel Boulud, Eric Ripert, and, regarding the years in Las Vegas, Tony Marnell, Michael Demers, and Michael Jordan.

I am grateful for the openness with which my questions on many subjects were met with over the years by the late Charlie Trotter, Billy Durney, Mike Conlon, Shamus Jones, Garrett Oliver, Howard Silberstein, Brian Fredericksen, Michael London, Ed Espe Brown, Margaret Greenwood, Rick Nahmias, Luis Yepiz, Ernest Miller, Sarah Spitz, Robert Gohstand, Dylan Roby, and Alex Ortega. Harold McGee kindly shared some of the publishing history of *On Food and Cooking,* and chefs Mourad Lahlou, Jeremy Fox, Josef Centeno, and Carlos Salgado patiently discussed culinary influences and their early days as chefs.

The book would not exist without Dan Halpern's faith in it. I am grateful for his continuing belief in the project and for the team at Ecco, especially Eleanor Kriseman. Her vision of what the book could be encouraged my work, and her skillful editing brought form to the story and clarity to the page. It has been a pleasure working with her.

Several friends have been particularly generous with insight. I am grateful to Margot Dougherty, Robert Wemischner, Amy Albert, Michael Mullen, Steve Oney, Sylvia Tan, Russ Parsons, Providence Cicero, Rick Nelson, Tom Sietsema, Jonathan Sanoff, Alain Giraud, and Gerald Hirigoyen. My mother-in-law, Lila Nadell, kept me supplied with interesting clippings, and my father-in-law, Eddie, was always ready to accompany me to the farthest farm to conduct an interview. My parents, Helen Sullivan and Michael Kuh, supported the project from the start. Bryony Harper and my sister, Micaela, provided much encouragement. My wife, Bonnie, and our children, Simon and Isabel, have been great travel partners as I undertook far-flung research, and I hope the fun we had on those trips made up for my absence during the writing.